Lecture Notes in Mathematics

A collection of informal reports and seminars
Edited by A. Dold, Heidelberg and B. Eckmann, Zürich

95

A. S. Troelstra

University of Amsterdam, Amsterd

Principles of Intuitionism

Lectures presented at the summer conference on Intuitionism and Proof theory (1968) at SUNY at Buffalo, N.Y.

1969

Springer-Verlag Berlin · Heidelberg · New York

Library of Congress Catalog Card Number 74-88182 Printed in Germany. Title No. 3701

Contents

§ 1. Introduction

1.1

This paper is intended as an introduction to the principles of intuitionism. Basic notions, not formal systems are emphasized; proof theoretic results are mentioned to illustrate relative power and formal consequences of various principles. In other words, this paper presents the material needed to recognize certain formal systems as representing fragments of intuitionistic mathematics. The application of various principles in mathematics is illustrated by suitable examples.

Although the paper has more or less a survey character, it is not exhaustive; various important subjects are summarily treated or mentioned only (e.g. Gödel's Dialectica interpretation and functionals of higher type, completeness problems of intuitionistic logic).

The selection of the examples from mathematics was determined by their usefulness as illustrations and by personal preference; thus many well-developed subjects of intuitionistic mathematics are not touched upon (e.g. measure theory, algebra, projective geometry). For information regarding these subjects the reader is referred to Heyting's introduction [H 1966].

As regards presentation of the subjects, there is indebtedness to many sources. I wish to mention especially Professor Kreisel's lectures on intuitionism in Stanford (1966-1967) and Professor Heyting's lectures on intuitionism in Amsterdam (1960-61).

1.2

The subject of intuitionism might be described, at a first try, as "constructive mathematical thought".

This succinct description immediately raises a problem: what do we mean by "constructive"? I might try to answer this question by a mathematical definition, which would select from the whole universe of possible mathematical arguments the constructive ones. However, there are two disadvantages to this procedure, even if I were to succeed in giving such a definition in a formally satisfactory way. Firstly, this would not lead to an autonomous development of constructive mathematics, independent of any reference to "all possible mathematics"; and secondly, we would be left with the (if you want, extra-mathematical, but not irrelevant) question of whether there exists a basic intuitive concept of constructive mathematics corresponding to the formal definition.

On the contrary, I want to start from the idea that there is a legitimate distinction between constructive and non-constructive mathematical thought. In other words: in some cases we feel no doubt about some argument being constructive, whereas in other cases we feel clearly that the argument as presented to us does not have the form of a construction.

Let us consider a simple example. Natural numbers may be regarded as constructions of a very simple kind. If I express Fermat's last theorem by

$$F \quad \begin{cases} \text{For all integers } n > 2\text{, and all positive integers} \\ x, y, z : x^n + y^n \neq z^n \end{cases}$$

then classically we accept a statement like

$$m = 0 \text{ if } F \text{ holds, } m = 1 \text{ otherwise} \qquad (1)$$

as determining a natural number m. From a constructive point of view we cannot assert that (1) determines m, since we have no way of deciding F.

In dealing with our subject, we shall admit some idealizations of reality (you might as well say simplifications) in order to get anywhere at all. We shall point out some of these simplifications as we proceed.

For example: although we have a much clearer concept of a construction representing a small natural number like 3, than we have of a construction representing say 9^{9^9}, we do not make any distinction between them. Every natural number is constructive to the same degree, so to speak. It can be worthwhile, however, to make a distinction between various parts of intuitionistic mathematics relative to the degree of abstraction and idealization involved. I shall not attempt to do this systematically.

Two remarks have to be added in order to delimit the subject of intuitionism more accurately.

First. We may start with very simple concrete constructions, such as the natural numbers, and then gradually build up more complicated, but nevertheless "concrete" or "visualizable" structures. Finitism is concerned with such constructions only ([Kr 1965], 3.4). In intuitionism, we also want to exploit the idea that there is an intuitive concept of "constructive", by reflection on the properties of "constructions which are implicitely involved in the concept". (I.e. we attempt to discover principles by introspection.)

Finitist constructions build up "from below"; reflecting on the general notion represents, so to speak, an approach "from the outside", "from above". Logic presents the simplest (and in its kind rather elementary) example of the approach from the outside.

The approach "from the outside" leads us to considering constructions applied to (arbitrary) constructions applied to constructions ... etc. For these reasons, intuitionism might be termed "abstract" in constrast to finitism as being "concrete".

From the preceding remarks it will be clear that we certainly cannot expect the conceptual basis of intuitionism to be simpler than the conceptual basis of classical mathematics.

Second. The (mental) constructions we consider, are thought of as to exist in the mind of an individual (idealized) mathematician. The language of mathematics is an attempt (necessarily nearly always inadequate) to describe these mental constructions. Talking about intuitionistic mathematics is therefore a matter of suggesting analogous mental constructions to other people. Similarity between the thought processes of various human individuals makes such communication possible.

This mathematician, who occupies himself with constructive mathematics, is an idealized creature; his ideas are supposed to be clear and distinct, not hazy and confused, as ours often are.

For the mathematicians of real life, mathematical language is an important help in keeping track of their thought and in sorting out confusion; but language is not essential to the idealized mathematician.

For an intuitionist, mathematics is not to be found in formal systems, as it is for the formalist. Nevertheless, formalization is a very important tool in our research of intuitionistic mathematics, for checking the principles used in the proof of a theorem, to suggest our constructions to others, and as a shorthand. By these functions formalization helps us in making our intuitive, languageless insights more precise. But if we want to study intuitionism, formal systems are not to be separated from their interpretation.

Finally, a word of warning. Intuitionism is a complicated and often tricky subject. Many topics have not yet been investigated sufficiently, therefore the subject is in a far from finished state. Accordingly, some considerations in these lectures are in need of correction or a more accurate formulation or a further development. But the paper will have served its purpose if the reader has got an idea of the program of intuitionism.

§ 2. Logic

2.1

Logic represents an example of the approach "from the outside": reflection on general principles about constructions and constructive proofs.

Logic is elementary in the sense that it does not make use of deep insights about the structure of constructive proofs; on the other hand, it is also very non-elementary as regards interpretation, and in a sense, highly impredicative.

The "elementary" character of logic is illustrated by the fact that e.g. intuitionistic propositional logic is capable of so many different interpretations (which means that the insights about constructions used in the interpretation of logic are not specific, since so many other interpretations are possible).

Let me first present a rough description of the meaning of the logical constants, which is all you need to understand the other sections.

A proof of $A \vee B$ is given by presenting a proof of A or a proof of B.

A proof of $A \wedge B$ is given by presenting a proof of A and a proof of B.

A proof of $A \rightarrow B$ is given by a construction which transforms any given proof of A into a proof of B, together with a proof of this fact.

$\neg A$ is proved by giving a proof of something like $A \rightarrow 1 = 0$.

$\bigvee x\, Ax$ is proved by exhibiting a construction c and a proof d which proves Ac.

$\bigwedge x\, Ax$ is proved by giving a construction scheme which yields a proof of Ax_0 for any specific x_0, together with a proof of this fact. In general, the quantifiers are supposed to range over a domain which has been "grasped as a whole" previously.

2.2

We shall try to make these explanations more precise in terms of Kreisel's "meaning functions" ([Kr 1965]). The following introduction to the theory of constructions is only a rough sketch; see subsection 2.3.

We distinguish between <u>constructions</u> and general notions, in short: <u>notions.</u> Constructions are the objects of our mathematical research; proofs are considered to be constructions.

Notions are decidable properties of constructions. We cannot identify notions with constructions, this would lead us immediately into paradoxes of denotation and self-reference (for an example see [Kr 1965], 2.152). One may compare this with the rôle of classes relative to sets in NBG-set theory. Proofs may use notions; herein is the impredicative character of our explanations. The domain of all constructions is not "given" or "a priori bounded", does not "exist" as an entity; and notions are properties which extend "automatically" with every extension of the domain of constructions.

In this section, we use lower case letters a, a', a", ..., b, b', ..., c, c', ..., d, ... for constructions, x, y, z, ... as dummy variables for constructions. Greek lower case letters α, β, ..., π, π_A, ... are used for notions; a notion α is interpreted as a function into $\{0,1\}$ (characteristic function) such that

$$\alpha c = 0 \quad \text{iff} \quad c \text{ has the property } \alpha.$$

When a, b are constructions, then $c = \langle a,b \rangle$, the pair of constructions a,b is also a construction. Conversely we define c_1, c_2 by: $c_1 = c_2 = c$ if c is not a pair, $c_1 = a$, $c_2 = b$ if $c = \langle a,b \rangle$.

A construction may consist of a schema applicable to other constructions. Our first idealization is included in the following assumption:

" a is applicable to b " is a notion (i.e. a decidable property). (1)

Hence we may suppose an application operator $\cdot(\cdot)$ to be defined by stipulating $a(b) = a$ say, if a is not applicable to b, and $a(b) = c$ if a applied to b yields c.
Another basic assumption (and idealization) is given by the assumption

" c is a proof of A " is a notion for any given assertion A. (2)

More general

" c is a proof of $A(c_1,\ldots, c_n)$ " is a notion for any given predicate $A(x_1,\ldots, x_n)$. (2')

The notion associated with A according to (2) will be indicated by π_A:

$$\pi_A c = 0 \quad \text{iff} \quad c \text{ is a proof of } A.$$

Likewise for $A(x_1, \ldots, x_n)$

$$\pi_A(c;\ c_1, \ldots, c_n) = 0 \text{ iff } c \text{ is a proof of } A(c_1, \ldots, c_n).$$

If we keep to our purely subjective point of view, (2) and (2') are rather natural: if we are in doubt whether a construction c proves A, then apparently c does not prove A for us!

Compare this with the provability predicate in formal systems: it is decidable whether a certain ordered sequence of formulae is a proof of a given formula or not.

Logical operators are operations which transform meaning functions into new meaning functions (of compound assertions). The simplest cases are $\vee$ and $\wedge$:

Conjunction:

$\pi_{A \wedge B}(c) = 0$ iff $c = \langle c_1, c_2 \rangle$ and $\pi_A(c_1) = 0$ and $\pi_B(c_2) = 0$

or more simply, since c_1, c_2 are always defined:

$$\pi_{A \wedge B}(c) = 0 \text{ iff } \pi_A(c_1) = 0 \text{ and } \pi_B(c_2) = 0.$$

Disjunction:

$\pi_{A \vee B}(c) = \pi_A(c) \cdot \pi_B(c)$; i.e. $\pi_{A \vee B}(c) = 0$ iff $\pi_A(c) = 0$ or $\pi_B(c) = 0$.

In the definition of $\pi_{A \wedge B}$, $\pi_{A \vee B}$ we meet the words "and" and "or", but their use is quite elementary, applied to decidable relations.

For the construction of $\pi_{A \to B}$ we need a new basic principle:

" c is a proof with a free variable d of $\alpha(d) = 0$ "
is a notion for any notion α. (3)

In other words, we are able to recognize if a construction c is a universally applicable schema such that c applied to d proves $\alpha(d) = 0$. We introduce therefore a notion π:

$$\pi\ (c, \lambda d. \alpha d = 0) = 0 \text{ iff } c \text{ is a proof with free variable } d \text{ of } \alpha d = 0.$$

Now we state

Implication:

$\pi_{A\to B}(c) = 0$ iff $c = \langle c_1, c_2 \rangle$ and

$$\pi(c_1, \lambda d.\ (1 \dot{-} \pi_A d)\pi_B c_2 d = 0) = 0.$$

($\dot{-}$ denotes ordinary cut-off subtraction.)

$x \supset y$ (classical implication) will be used for $x = 0 \to y = 0$.

Negation:

$\pi_{\neg A}(c) = 0$ iff $\pi(c, \lambda d.\ (1 \dot{-} \pi_A d) = 0) = 0$.

Universal quantification:

Let $A(x_0, \ldots, x_k)$ be a predicate.

$\pi_{\bigwedge x_0 A}(c;\ c_1, \ldots, c_k) = 0$ iff $c = \langle d_1, d_2 \rangle$ and

$$\pi(d_1(c_1, \ldots, c_k), \lambda d.\pi_A(d_2(d, c_1, \ldots, c_k);\ d, c_1, \ldots, c_k) = 0) = 0.$$

Existential quantification:

$\pi_{\bigvee x_0 A}(c; c_1, \ldots, c_k) = 0$ iff $c = \langle d_1, d_2 \rangle$ and $\pi_A(d_1; d_2, c_1, \ldots, c_k) = 0$.

The restriction of logic to $\to, \wedge, \vee, \neg, \bigwedge, \bigvee$ is more or less determined by tradition; other operations on meaning functions are conceivable, which are not definable in terms of $\to, \wedge, \vee, \neg, \bigwedge, \bigvee$.

Logical validity for the intuitionistic propositional calculus might be expressed as

$$F(P_1, \ldots, P_k) \text{ is intuitionistically valid iff}$$

$$\bigwedge \pi_{P_1} \ldots \bigwedge \pi_{P_k} \bigvee c\ \pi_F(c) = 0. \qquad (4)$$

($P_1, \ldots, P_k$ are propositional variables.)

Let $F(R_1, \ldots, R_k)$ denote a formula of predicate logic with predicate variables $R_1, \ldots, R_k$, R_i a predicate with ρ_i arguments; suppose F contains dummy variables $x_0, \ldots, x_n$. Then F is said to be valid if $\bigwedge x_0 \ldots \bigwedge x_n F$ is valid.

The validity of closed formulae of the predicate calculus is also defined by (4), if $P_1, \ldots, P_k$ are taken to be predicate variables.

As examples, we discuss two logical theorems.

2.2.1. Theorem. $A \to \neg\neg A$ is intuitionistically valid.
We have to show $\bigwedge \pi_A \bigvee a(\pi_{A\to\neg\neg A}(a) = 0)$.

Proof. Let

$$\pi_A(b) = 0 \quad \text{(hypothesis)}. \tag{5}$$

$\pi_{\neg A}(c) = 0$ iff $\pi(c, \lambda d.\pi_A d = 1)$. Hence by our hypothesis

$$\pi_{\neg A}(c) = 1. \tag{6}$$

The preceding argument represents a proof $a'(b)$ with free variable c of (6), i.e.

$$\pi_{\neg\neg A}(a'(b)) = 0. \tag{7}$$

Therefore

$$\pi_A(b) = 0 \supset \pi_{\neg\neg A}(a'(b)) = 0 \quad \text{(elimination hypothesis)}. \tag{8}$$

(5) - (8) represent a proof c' with free variable b of (8). Therefore

$$\pi_{A\to\neg\neg A}\langle c', a'\rangle = 0.$$

2.2.2. Theorem. $\neg \bigvee x\, Ax \to \bigwedge x \neg Ax$ is intuitionistically valid.

Proof. We look for a construction a such that

$$\pi_{\neg\bigvee xA \to \bigwedge x\neg A}(a) = 0. \tag{9}$$

Let

$$\pi_{\neg\bigvee xA}(b) = 0 \quad \text{(hypothesis)} \tag{10}$$

then

$$\pi(b, \lambda c.(1 \dot{-} \pi_{\bigvee xAx}(c)) = 0) = 0 \tag{11}$$

By (11)

$$\pi_{\bigvee xA}\langle d, d'\rangle = 1 \tag{12}$$

for arbitrary d, d', so

$$\pi_A(d;d') = 1. \tag{13}$$

(10) - (13) is a proof $a'\langle b, d'\rangle$ with free variable d of (13) under hypothesis (10), hence

$$\pi_{\neg A}(a'\langle b, d'\rangle; d') = 0. \tag{14}$$

(10) - (14) represents a proof $b'(b)$ with free variable d' of (14), hence if

we take $c'(b) = \langle b'(b), a''(b) \rangle$ and $a''(b)(d') = a'\langle b,d'\rangle$ for all d', we have

$$\pi_{\wedge x \neg A}(c'(b)) = 0. \quad (15)$$

(10) - (15) is a proof b'' with free variable b of

$$\pi_{\neg \vee x A}(b) = 0 \supset \pi_{\wedge x \neg A}(c'(b)) = 0 \text{ (elimination hyp.)}. \quad (16)$$

Hence

$$\pi_{\neg \vee x A \to \wedge x \neg A}\langle b'', c' \rangle = 0. \quad (17)$$

Rules are treated as follows. If we want to show say

$$\frac{F, G \vdash H}{F', G' \vdash H'}$$

then we must show that to every b such that $\pi_{F \wedge G \to H}(b) = 0$ we can find a c such that $\pi_{F' \wedge G' \to H'}(c) = 0$.

2.3

Intentionally I refrained from a systematic development of the theory of constructions; the subject is far from being in a finished state. The preceding rough sketch of a general theory of constructions in which we can explain the meaning of the logical constants is in need of refinement. For a detailed development of the theory of constructions, including intuitionistic arithmetic, we refer to [Gn 1968], [Gn].

The main refinements in Goodman's development in [Gn 1968], [Gn] as compared with the sketch given above are 1°) rules, functions need not to be defined everywhere; and most basic point of all: 2°) proofs must be about objects <u>already constructed</u>, and likewise quantification is supposed to be over domains which are already "understood", had been "grasped" before. This automatically introduces levels; the lowest level consists of constructive rules operating on each other.

For any given level L the proof-predicate: d proves $cx = 0$ for all $x \in L$ is decidable and belongs to the next higher level. Each level is universally decidable in the sense of 5.3. The general idea might be summarized by saying: "there is an a priori bound on the possible proofs for any definite assertion".

2.4

Technical results about intuitionistic logic are well-known and easily available in the literature. For a Hilbert-type system for intuitionistic logic see e.g. [K 1952], Ch. IV. A system with Gentzen-type rules, for which cut-elimination can be proved is to be found in [K 1952], § 77.

For future reference we mention a convenient system essentially due to Spector ([Sp 1962]).

<u>Propositional axioms and rules</u>: (I) $P \to P$; (II) If Q, then $P \to Q$; (III) If P and $P \to Q$, then Q; (IV) If $P \to Q$ and $Q \to R$, then $P \to R$; (V) $P \wedge Q \to P$ and $Q \wedge P \to P$; (VI) $P \to P \vee Q$ and $Q \to P \vee Q$; (VII) If $P \to R$ and $Q \to R$, then $P \vee Q \to R$; (VIII) If $R \to P$ and $R \to Q$ then $R \to P \wedge Q$; (IX) If $P \wedge Q \to R$, then $P \to (Q \to R)$; (X) If $P \to (Q \to R)$ then $(P \wedge Q) \to R$; (XI) $\perp \to P$.

Here $\perp$ stands for a contradiction, "falsehood". In accordance with our general interpretation of the logical constants P may be interpreted as $P \to \perp$. If we want to avoid a separate symbol for falsehood, we must replace (XI) e.g. by:

(XI)' $(A \wedge \neg A) \to B$; (XI)" $\neg B \to (B \to (A \wedge \neg A))$; (XI)"' $(B \to (A \wedge \neg A)) \to \neg B$.

<u>Axioms and rules for quantifiers</u>:

Let x be a variable not occurring free in Q, and let t be a term free for x in Px. Then

(XII) If $Q \to Px$, then $Q \to \bigwedge x Px$; (XIII) If $Px \to Q$, then $\bigvee x Px \to Q$;
(XIV) $\bigwedge x Px \to Pt$; (XV) $Pt \to \bigvee x Px$.

Axioms and rules for equality as usual.

Intuitionistic propositional logic is decidable ([K 1952], § 80); for intuitionistic predicate logic the interpolation theorem is provable (see [S 1962], [N 1966]). The completeness of intuitionistic predicate logic for the <u>intuitionistic</u> interpretation is discussed in [Kr 1958], [Kr 1958A], [Kr 1962]. We remark here that completeness for intuitionistic propositional logic is proved fairly easily ([Kr 1958]).

§ 3. Elementary arithmetic

3.1

Arithmetic may be regarded as a piece of intuitionistic mathematics which to a large extent may be approached "from below", starting from very simple elementary constructions.

Natural numbers are conceived as constructions of a very simple kind, obtained by juxtaposing units. The basis of this concept is the observation that we can conceive a unit, then another unit, look upon this two-ity (pair) as a new entity, and repeat this process as often as we like. In a picture simply

I II III IIII, ...

These very simple constructions are so to speak their own proof: for the concept of a certain natural number, the proof that it is a natural number is given by the number itself, because its mode of generation is at the same time "proof" that it has been obtained by this process of generation of natural numbers.

The notion of a successor x^+ of a number x is clear, and also the properties

$$\left.\begin{array}{l} 0 \neq x^+ \\ x=y \leftrightarrow x^+ = y^+ \end{array}\right\} \qquad (1).$$

Let Q be any property of natural numbers. The induction principle states (x, y, z variables for natural numbers)

$$Q0 \wedge \bigwedge x(Qx \rightarrow Qx^+) \rightarrow \bigwedge x\, Qx \qquad (2).$$

The justification of this principle may be given as follows.
Take any natural number y, and let $Q0$, $\bigwedge x\,(Qx \rightarrow Qx^+)$.
Now we construct proofs of $Q0, \ldots, Qy$ parallel to the generation of $0, 0^+, 0^{++}, \ldots$:

$$\begin{array}{lll} \cdot & Q0, & \\ Q0, & Q0 \rightarrow Q1 & \vdash Q1, \\ Q1, & Q1 \rightarrow Q2 & \vdash Q2, \\ & \ldots & \\ Q(y-1), & Q(y-1) \rightarrow Qy & \vdash Qy. \end{array}$$

Thus , by parallelling the construction of natural numbers, we prove $\bigwedge y\, Qy$.

We have been rather explicit about this point, since an analogous argument occurs in a much more complicated situation later on.

The existence of primitive recursive functions is also seen by a "step by step" construction of function values.

3.2

A formal system for intuitionistic first order arithmetic may be found e.g. in [K 1952], Ch. IV or in [Sp 1962]. The new axioms are the axioms for $^{+}$ (formulas (1)), recursion equations for $+, \cdot$, and induction (2) with respect to Q formulated in the language of first order predicate logic with equality and $^{+}, +, \cdot$.

For future reference, we state the following important result.

3.2.1. Theorem (Gödel; see [K 1952], § 81). Let $\mathcal{L}$ denote intuitionistic propositional logic, predicate logic or first order arithmetic; let $\mathcal{L}^{+}$ denote the corresponding classical system. We define a translation as follows:

$$
\begin{array}{lll}
(a) & P^{-} & = \neg\neg P \quad \text{(for prime formulae } P) \\
(b) & (P \to Q)^{-} & = P^{-} \to Q^{-} \\
(c) & (P \wedge Q)^{-} & = P^{-} \wedge Q^{-} \\
(d) & (\neg P)^{-} & = \neg P^{-} \\
(e) & (P \vee Q)^{-} & = \neg(\neg P^{-} \wedge \neg Q^{-}) \\
(f) & (\bigwedge x Px)^{-} & = \bigwedge x P^{-}x \\
(g) & (\bigvee x Px)^{-} & = \neg \bigwedge x \neg P^{-}x .
\end{array}
$$

(Instead of (b) we may also take $(P \to Q)^{-} = \neg(P^{-} \wedge \neg Q^{-})$).

Then $P \in \mathcal{L}^{+}$ iff $P^{-} \in \mathcal{L}$.

3.2.2. Remark. In the case of arithmetic we can take $P^{-} = P$ for prime formulae, since the prime formulae of arithmetic are decidable. (In the formal system this has to be proved by induction from $0 \neq x^{+}$, since we do not have the excluded third.)

The theorem yields a consistency proof of classical arithmetic relative to intuitionistic arithmetic. Moreover, it shows that $\mathcal{L}$ is not poorer than $\mathcal{L}^{+}$.

3.2.3. Notation. We use the functions sg, $\dot{-}$, etc. as defined in [K 1952]. N denotes the collection of natural numbers.

§ 4. Species

4.1

In this section we introduce the notion of a species. Species may be regarded as intuitionistic analogues of the classical sets. Roughly speaking, species are properties which are in turn considered to be mathematical objects (entities).

Suppose that we have any well defined collection A of mathematical entities (such that the collection itself may be considered to be an object, e.g. $A = N$).

Then well-defined properties of elements of A are considered to be species (intuitionistic form of the comprehension principle). The extent of the notion of species is therefore determined by the interpretation of "well-defined". But it is clear anyway that "well-defined" for a property P implies that we know what constitutes a proof of Px for an $x \in A$.

It will be clear that an element x of A can only be admitted as an element or member of P ($x \in P$) if x has been or might have been defined before (independently of) P.

In terms of the theory of constructions (as used in § 2 to explain the intuitionistic interpretation of the logical constants) species correspond to notions which themselves are constructions. In other words, a predicate A with a meaning function π_A is a species if there exists a construction c such that

$$\pi_A(a;b) = 0 \leftrightarrow c \langle a,b \rangle = 0$$

for all constructions a,b.

Essentially impredicative applications of the comprehension principle involve quantifications over all subspecies of the basic collection A; and if we admit e.g. universal quantification over all subspecies of A ("subspecies" is defined like "subset") there seems to be no reason not to accept $\mathfrak{P}(A) = \{X : X \subseteq A\}$ (the power-species) as a species.

Typically predicative applications of the comprehension principle involve quantifications over elements of the basic collection A only.

The theory of species is a rather underdeveloped domain of intuitionistic mathematics, and Brouwer's comments are scarce. I introduce the general notion of a species at this stage mainly for the purpose of a convenient formulation of certain results in the sequel. The discussion of the notion is resumed in § 15. For the time being, we shall leave open the question as to the extent to which properties are admitted as species. We remark, however, that there seems little objection against accepting all essentially predicative applications of the comprehension principle; therefore we accept e.g. all arithmetical predicates of the natural numbers as defining species.

Most of the applications of the comprehension principle in the sequel are non-essential and predicative relative to the assumption that some basic properties like those of being a natural number or a lawlike funtion from N into N (see § 5) are species. Notable exceptions are K in § 9 and WO in § 14. It is clear that any use of "species" where one might just as well say "unary predicate" is non-essential.

4.2.

In classical set theory, sets are determined by their elements. In agreement with this approach we may define equality between species X, Y by:

$$X = Y \longleftrightarrow \bigwedge x(x \in X \longleftrightarrow x \in Y).$$

However, it is necessary to stress here the intuitionistically very relevant distinction between intensional and extensional equality.

Our creation of mathematical objects starts with concrete, immediately given constructions, like natural numbers, or descriptions of mappings; between such constructions we have a definitional equality (say $\equiv$) which is decidable; $x \equiv y$ indicates that x,y are <u>given</u> to us as the same object, "thought". Let us call such entities objects of order zero. Properties of objects of order zero, e.g. properties of natural numbers, serve to introduce species; the definition of such an object is itself an object of type zero; but the "extension of the property", i.e. the property as related to extensional equality, might be termed an object of type 1. Likewise we may form objects of type 2, 3, ... etc.; objects of type n + 1 are species of objects of type n, in relation with extensional equality. The property which defines a species of order n + 1 may itself be looked upon as an object of type zero.

4.2.1. <u>Definitions</u>. A species X is said to be <u>inhabited</u> or <u>secured</u> if $\bigvee x(x \in X)$. Y is <u>detachable</u> in X, if $\bigwedge x \in X(x \in Y \vee x \notin Y)$.

4.2.2. <u>Definition</u>. A binary relation $\#$ on a species X is said to be an apartness relation on X, if for all $x,y,z \in X$

(a) $\neg x \# y \longleftrightarrow x = y$

(b) $x \# y \longrightarrow y \# x$

(c) $x \# y \longrightarrow x \# z \vee z \# y$.

4.2.3. Remark. Equality on a species with an apartness relation is stable, i.e. $\neg\neg x = y \rightarrow x = y$, since

$$x = y \leftrightarrow \neg x \# y \quad \text{and} \quad \neg\neg\neg x \# y \leftrightarrow \neg x \# y.$$

4.2.4. $X \times Y$, X^2, X^3, $X \cup Y$, $X \cap Y$, $X - Y$, $X \subseteq Y$ are defined in the usual way.

§ 5. Sequences and constructive (lawlike) objects

5.1

5.1.1. Definition. A mapping φ from a species X into a species Y is any kind of process which assigns to any $x \in X$ an $y \in Y$, and such that $x = x' \rightarrow \varphi x = \varphi x'$. (This stipulation is necessary whenever $=$ does not denote basic definitional equality, e.g. in the case where the elements of X are themselves species.) A mapping φ from X into Y is said to be of type $(X)Y$ (also $\varphi \in (X)Y$). A mapping $\varphi \in (X)Y$ is said to be bi-unique (an injection) if

$$\bigwedge x \in X \bigwedge x' \in X \, (\varphi x = \varphi x' \rightarrow x = x').$$

φ is said to be weakly bi-unique (weak injection) if

$$\bigwedge x \in X \bigwedge x' \in X \, (x \neq x' \rightarrow \varphi x \neq \varphi x').$$

Remarks. In case the equality in Y is stable, i.e. $\neg\neg y = y' \rightarrow y = y'$, a weak injection is an injection. An injection $\varphi \in (X)Y$ possesses an inverse $\varphi^{-1} \in (\varphi[X])X$ $\quad (\varphi[X] = \{\varphi x : x \in X\})$.

5.1.2. Definition. Let X, Y be species with apartness relations $\#$, $\#'$ respectively. $\varphi \in (X)Y$ is said to be strongly bi-unique (a strong injection) if $\bigwedge x \in X \bigwedge x' \in X \, (x \# x' \rightarrow \varphi x \#' \varphi x')$.

5.1.3. Remark. In case X is a species with apartness relation $\#$, and φ is a bi-jection, $\varphi \in (X)Y$, $[X] = Y$, then an apartness relation $\#'$ on Y is defined by

$$x \#' y \equiv_D \varphi^{-1} x \# \varphi^{-1} y.$$

5.2

Integers and rationals are constructed from natural numbers in the same way as in classical mathematics. But in order to develop a theory of real numbers, we have to introduce the notion of a sequence.

In general, a sequence is a mapping of type $(N)X$, i.e. a process which associates with every natural number a mathematical object belonging to a certain species X.

We shall use $\chi, \chi', \chi'', \ldots$ for sequences in general. The general idea of a sequence leaves open various possibilities for further specialization. The simplest example is that of a constructive or lawlike sequence, which might be described as a sequence which is completely fixed in advance by a law, i.e. a prescription (algorithm) which tells us how to find for any $n \in N$ the nth member of the sequence. As typical examples one may think of say primitive recursive functions.

A lawlike sequence is therefore given to us by an algorithm together with a proof that the algorithm applied to any natural number produces a natural number, or more generally, an element of a species X. (It may be maintained that an algorithm when described in full also contains the proof of its own universal applicability.)

Lawlike sequences are considered to be definitionally equal (intensionally equal) if they are given to us in the same way. It will be clear that extensional equality $\bigwedge x(\chi x = \chi' x)$ does not imply intensional equality $(\chi \equiv \chi')$ for lawlike χ, χ'.

From now on we shall use a, b, c, d for lawlike sequences of (N)N (unless the context clearly indicates otherwise). We use $\bar{\chi}x$ for the finite sequence $\chi 0, \ldots, \chi(x-1)$; $\bar{\chi}0 = \emptyset$.

A hypothesis relative to lawlike sequences of (N)N, sometimes called "Church's thesis" or "the intuitionistic form of Church's thesis" states that every lawlike function of (N)N is recursive, or in Kleene's symbolism

$$\bigwedge a \bigvee n \bigwedge x \ (ax = \{n\}x) \qquad (1)$$

($\{n\}$ recursive function with Gödel number n). Compare [Kr 1965] 2.72.

Many formal systems proposed for fragments of intuitionistic mathematics turn out to be consistent with (1). See also § 16.3.

Another (fairly simple) notion, almost the opposite of a lawlike sequence, is that of a lawless sequence (of (N)N). Such a sequence is conceived as a source of values, such that at any stage we know a finite sequence of values $\chi 0, \chi 1, \ldots, \chi n$, while nothing is known about future values.

Compare this with the casts of a die; at any moment, finitely many casts are known, but about future casts nothing can be said.

One may also think of lawless sequences as objects abtained by "abstracting" from anything but what is needed to ensure that to every x a value χx can be found. (E.g. one may create a lawless sequence by thinking of say $\lambda x.0 \equiv a$ as a process which generates 0, 0, 0, ...; in applying at any stage any operation to this sequence we do not use the law, only the initial segment 0, 0, ..., 0 obtained at that stage. 0, 0, 0, ... regarded as a lawless sequence is such that at any stage we conceive all possible values at further arguments as possible.) We return to lawless sequences in § 9.

Lawlike sequences and natural numbers provide us with the simplest examples of complete objects, i.e. mathematical objects which at a certain stage are completely described.

Lawless sequences are the simplest examples of "incomplete" objects, objects which are being generated; at no stage the process is thought of as being finished.

The motivation for studying notions of sequence where the idea of determination by a law is abandoned, is provided by the circumstance that most of the operations on sequences of rational numbers that play a rôle in analysis, do not depend on the assumption of the sequences to be lawlike.

It will turn out, however, that lawless sequences, although interesting on their own account, and useful in the discussion of intuitionistic logic, are not very well suited to the development of a theory of real numbers and real-valued functions.

For a more satisfactory theory of the continuum and the real-valued functions we need a more complex notion, intermediate between lawless and lawlike; this notion of choice sequence will be discussed in § 10.

We may generalize the notion of a lawlike sequence to lawlike operations $\Phi \in (X)Y$; $\Phi \in (X)Y$ is called lawlike if Φ can be given completely by a description.

For those readers acquainted with the theory of creative subject, I remark here that the description of "lawlike sequence" does not necessarily exclude reference to the creative subject.

5.3

Let X be a (universally) decidable species with an intensional equality relation. (By "universally decidable" we mean that for any given construction it can be decided whether this construction represents an element of X or not. In particular, we shall suppose the species of natural numbers to be decidable.) Then we may formulate the following "form of the axiom of choice" or "selection principle": When A is an (extensional) predicate, then

$$\bigwedge x \in X \ \bigvee y \in Y \ A(x,y) \rightarrow \bigvee \Phi \in (X)Y \ \bigwedge x \ A(x, \Phi x) \qquad (2).$$

For if we have a proof of

$$\bigwedge x \in X \bigvee y \in Y \ A(x,y)$$

then this proof must contain a complete description (possibly involving parameters occurring in A) of an operation which assigns to every x a $y \in Y$.

When A does not contain non-lawlike variables, then Φ may be supposed to be lawlike.

If X is a species with an equality relation $=$, we may associate with X a species X' of "definitions of elements of X" (or in other words, consisting of the elements of X, but with definitional equality $\equiv$). Such a "definition of an element of X" may be thought of as a construction together with a proof that this construction represents an element of X. In general agreement with the principles described in section 2, if we suppose the equality of proofs and constructions to be decidable, it may be assumed that X' is universally decidable. If we have a proof of $\bigwedge x \in X \bigvee y \in Y \ A(x,y)$, the only conclusion that is immediately evident, is

$$\bigvee \Phi \in (X')Y \bigwedge x \in X' \ A(\overline{x}, \Phi x) \ (\text{where } \overline{x} = \{y : y \in X' \wedge y = x\}),$$

but there is no reason to suppose that a mapping of $(X')Y$ always can be extended to a mapping of $(X)Y$.

In case the equality on X is such that there exists a $\Psi \in (X')X'$ such that

$$\bigwedge x \in X' \bigwedge y \in X' \ (x = y \rightarrow \Psi x \equiv \Psi y \wedge \Psi x = x)$$

then $=$ is decidable, and (2) may be asserted, since it follows that

$$\bigwedge x \in X' \ A(\overline{x}, \Phi \Psi x),$$

for we have $A(\overline{\Psi x}, \Phi \Psi x)$, and also $\overline{\Psi x} = \overline{x}$; but A is extensional, i.e. $A(x,y) \wedge x = x' \rightarrow A(x',y)$, hence $A(\overline{x}, \Phi \Psi x)$.
$\Phi \Psi$ determines a $\Phi' \in (X)Y$ by $\Phi' \overline{x} = \Phi \Psi x$, hence $\bigvee \Phi' \bigwedge x \in X \ A(x, \Phi' x)$.

Important special cases of (2) may be formulated as follows (A not containing non-lawlike variables, $\{ \}$ denoting a pairing function of type $(N^2)N$):

$$\bigwedge n \bigvee \chi \ B(n, \chi, \chi_0, \ldots) \rightarrow \bigvee \chi' \bigwedge n \ B(n, \lambda m. \chi'\{n,m\}, \chi_0, \ldots) \qquad (3)$$

$$\bigwedge n \bigvee m \ B(n, m, \chi_0, \ldots) \rightarrow \bigvee \chi \ \bigwedge n \ B(n, \chi n, \chi_0, \ldots) \qquad (4)$$

$$\bigwedge n \bigvee a \ A(n,a) \rightarrow \bigvee b \ \bigwedge n \ A(n, \lambda m. b\{n,m\}) \qquad (5)$$

$$\bigwedge n \bigvee m \ A(n,m) \rightarrow \bigvee a \ \bigwedge n \ A(n, an) \qquad (6).$$

B in (3) and A in (5) are assumed to be extensional, i.e.

$$\chi = \xi \wedge B(n,\chi, \ldots) \rightarrow B(n,\xi, \ldots)$$
$$a = b \wedge A(n,a) \rightarrow A(n,b).$$

If we restrict χ to a specific notion of sequence, it remains to be seen whether the proofs of $\bigwedge n \bigvee \chi\, B$, $\bigwedge n \bigvee m\, B$ always provide a χ' of the same kind. We return to this point in § 10.

5.4

As will be clear from the preceding discussion, the extent of the notion of a law-like operator depends on our notion of constructive proof. Definitions by recursion are usually acceptable. More controversial in this respect is the discussion of bar recursion of higher type. For formal work concerning this principle see e.g. [Ho 1968]. The principle of bar recursion of lowest type may be justified by the discussions in sections 9, 10.

5.5

Let us return to the interpretation of the logical constants. Instead of using the general explanations of the logical constants described in § 2, there are alternative possibilities which may be explored.

For example, we might have looked upon the general explanation of the logical constants as a kind of "meta-explanation" outside intuitionistic mathematics.

Take e.g. a theorem of the general form $A \rightarrow B$. Our meta-explanation of "$\vdash A \rightarrow B$" tells us that we have a method which enables us to construct a proof of B from an arbitrary proof of A. The actual proof of the theorem "$A \rightarrow B$" then contains the mathematically precise formulation; the precise meaning of $\rightarrow$ <u>in this special case</u> is "explained" by the proof which explicitely presents the required method.

Likewise, a logical law is interpreted as a metamathematical schema to be applied in proofs.

In this manner, logic is not properly included in mathematics. Thus, we may go a long way without using the explanation of the logical operations in full generality. Freudenthal, in his paper [F 1936] has expressed this point of view as follows:

> "Each theorem, once it has been correctly formulated, contains its own proof. Intuitionistically speaking, a theorem (in the usual sense) is a short, preliminary orientation, a kind of summary, whereas the theorem proper is given by the proof."

Gödel's Dialectica interpretation [G 1958] is related to this approach. If we take a certain formal language suitable for describing a fragment of intuitionistic mathematics, we are at liberty to eliminate some of the troublesome logical operations (like implication) in favor of defined notions which are conceptually simpler and which "approximate" the operations which they replace.

In Gödel's paper, an essentially "logic free" interpretation of intuitionistic arithmetic is given, i.e. every arithmetical formula A has an interpretation of the form $\bigvee s \bigwedge t\ A'(x,t)$. s,t lawlike operations, A' decidable. It can be proved that if A is a formula of intuitionistic arithmetic, then $\bigvee s \bigwedge t\ A'(s,t)$ is provable in a logic free theory of lawlike operations, i.e. we can construct s in the theory such that there is a free variable proof (computational) of $A'(s,t)$ (w. r. t. the free variable t).

This result has been extended to analysis by C. Spector; see [Ho 1968] for an improved and smooth presentation.

The essential steps in the interpretation of implication in Gödel's theory occurs when, after replacing $\bigvee s \bigwedge t\ A(s,t) \to$ $\bigvee s' \bigwedge t'\ B(s',t')$ by $\bigwedge s(\bigwedge t\ A(s,t) \to \bigvee s' \bigwedge t'\ B(s',t'))$, this in turn is replaced by

$$\bigwedge s \bigvee s'(\bigwedge t\ A(s,t) \to \bigwedge t'\ B(s',t')),$$

and when, after replacing the previous formula by $\bigwedge s \bigvee s' \bigwedge t'(\bigwedge t\ A(s,t) \to B(s',t'))$, we replace this by

$$\bigwedge s \bigvee s' \bigwedge t' \bigvee t(A(s,t) \to B(s',t')).$$

The final step yields $\bigvee S' \bigwedge T \bigwedge s \bigwedge t'(A(s,Tst') \to B(S's,t'))$.
The resulting implication is essentially logic-free. This interpretation of the implication represents a strengthening relative to the original intuitionistic interpretation.

§ 6. Elementary theory of real numbers

6.1

In this section we develop a tiny part of the elementary theory of real numbers. We do not specify the notion of sequence we have in mind; to be definite, one may think of all sequences involved as being lawlike. The developments in this section are also valid for choice sequences and to a limited extent for lawless sequences.

We shall introduce real numbers by fundamental or Cauchy-sequences of rationals. The methods of nested intervals or Dedekind-cuts may also serve as a basis, but the approach via fundamental sequences is technically the simplest one.

We shall write $\langle x_n \rangle_{n=1}^{\infty}$ or shortly $\langle x_n \rangle_n$ for a sequence $x_1, x_2, \ldots$. x, y, z are used for reals in this section, r, s for rationals, k, n, m, i for natural numbers.

6.1.1. Definition. A sequence of rationals $\langle r_n \rangle_n$ is a real number generator if

$$\bigwedge k \bigvee n \bigwedge m (|r_n - r_{m+n}| < 2^{-k}).$$

Two real number generators $\langle r_n \rangle_n$, $\langle s_n \rangle_n$ are said to be equivalent (notation $\langle r_n \rangle_n \sim \langle s_n \rangle_n$) if

$$\bigwedge k \bigvee n \bigvee m (|r_{n+m} - s_{n+m}| < 2^{-k}).$$

Just as in the classical theory, one shows $\sim$ to be an equivalence relation. The equivalence classes are called real numbers. (If we think of lawlike real number generators, the corresponding equivalence classes might be called lawlike or constructive real numbers.)

It is evident how to define addition and multiplication for real number generators.

6.1.2. Definition.

$$\langle r_n \rangle_n + \langle s_n \rangle_n = \langle r_n + s_n \rangle_n$$

$$\langle r_n \rangle_n \cdot \langle s_n \rangle_n = \langle r_n . s_n \rangle_n$$

One proves easily

$$\langle r_n \rangle_n \sim \langle r'_n \rangle_n \rightarrow \langle r_n \rangle_n + \langle s_n \rangle_n \sim \langle r'_n \rangle_n + \langle s_n \rangle_n \quad \text{etc.}$$

exactly as in the classical theory. Hence for the reals we may stipulate

6.1.3. Definition. $\langle r_n \rangle_n \in x \wedge \langle s_n \rangle_n \in y \rightarrow \langle r_n + s_n \rangle_n \in x + y \wedge \langle r_n . s_n \rangle_n \in x . y$.

With the definition of x^{-1} we meet difficulties in connection with zero; this provides us with the occasion to expand on the subject of intuitionistic counter-examples.

6.2.

There seems to exist a widespread misunderstanding that "intuitionistic" counter-examples constitute an essential part of intuitionistic mathematics, and also that intuitionists like to give these examples for the fun of proving other people to be wrong. This popular belief is, I think, wrong on both counts.

But it is a fact that these counterexamples are often useful, if only to save us the trouble of attempting to mimick classical arguments when there is no reasonable hope of succeeding.

This is one reason for saying something about these counterexamples. The other reason is that they occur frequently in the literature.

Let me first state such an example in the conventional way. π denotes, as usual, the ratio between the circumference of a circle and its diameter. Let $\Pi_m(n)$ indicate the assertion: n is the number of the last decimal of the mth sequence of 10 consecutive numerals 7 in the decimal expansion of π. (I will not bother here to prove intuitionistically that π possesses such a decimal expansion; see [Br 1920]; or take e instead of π, and look at [H 1966], 2.3.)

It is a fact that up till now it is unknown whether

$$\bigvee n\ \Pi_1 n \vee \neg \bigvee n\ \Pi_1 n$$

and we do not have a method to settle this question.

Now we define a real number generator $\langle r_m \rangle_m$ by

$$\left.\begin{array}{ll} \neg \bigvee n \leq m\ (\Pi_1 n) \rightarrow r_m & = 0 \\ \Pi_1 n \wedge n \leq m \rightarrow r_m & = 2^{-n} \end{array}\right\}$$

$\langle r_m \rangle_m$ is a constructive real number generator, since the decimal expansion of π is evidently lawlike.

Let x_0 denote the real number defined by $\langle r_m \rangle_m$.
We have no method of deciding $x_0 = 0 \vee x_0 \neq 0$, since $x_0 = 0 \leftrightarrow \neg \bigvee n(\Pi_1 n)$, hence $(x_0 = 0 \vee x_0 \neq 0) \leftrightarrow (\bigvee n \Pi_1 n \vee \neg \bigvee n\ \Pi_1 n)$. x_0 might be called a floating number relative to zero.

We can state this in a more general way:
$\bigwedge x(x = 0 \vee x \neq 0)$ would imply for any predicate X of natural numbers:

$$\bigwedge n(Xn \vee \neg Xn) \rightarrow (\bigvee n\, Xn \vee \neg \bigvee n\, Xn)$$

for we can associate an x_X with every decidable X, like we did for Π_1.

It will be clear that these counterexamples do not provide mathematical refutations; they reduce a problem to unsolved problems of a certain kind, for which it is unlikely that we shall ever find a constructive solution. Specifically, we have no hope of proving $\bigwedge x(x = 0 \vee x \neq 0)$ unless we have a method of solving all problems $\bigvee n\, Xn \vee \neg \bigvee n\, Xn$ for decidable X. If x is supposed to be restricted to lawlike reals, then

$$\bigwedge x(x \neq 0 \vee x = 0) \rightarrow \bigwedge a(\bigvee n(an \neq 0) \vee \neg \bigvee n(an = 0)) \qquad (1)$$

since for any decidable X not containing free non-lawlike variables $\bigvee a \bigwedge n\,(an = 0 \leftrightarrow Xn)$ (§5).

In the literature the counterexamples mostly occur in the form of a reduction to a specific problem, but it is often worthwhile to bring them in a general form like (1).

For example, this can be useful for a comparison with recursion theory. So (1) would read, if we translate "lawlike" by "recursive": for recursive sets it is recursively decidable if a recursive set is empty or not (a is viewed as determining a characteristic function).

Reductions like (1) are especially relevant if we want to study formal systems with variables for constructive funtions for which we want to investigate if the "intuitionistic version of Church's thesis" may be added consistently.

The consequence of the counterexample discussed is, that we cannot be sure that x^{-1} is everywhere defined.

A few other examples of assertions for which we can give intuitionistic (weak-) counterexamples are

$\bigwedge x(x$ is rational or x is not rational),
$\bigwedge x(x$ possesses a decimal representation),
$\bigwedge x \bigwedge y(xy = 0 \rightarrow x = 0 \vee y = 0)$.

The actual construction of such counterexamples may be left as an exercise.

6.3.

6.3.1. Definition.

$$\langle r_n\rangle_n < \langle s_n\rangle_n \equiv_D \bigvee k \bigvee n \bigwedge m\ (s_{n+m} - r_{n+m} > 2^{-k}),$$

$$\langle r_n\rangle_n \# \langle s_n\rangle_n \equiv_D \bigvee k \bigvee n \bigwedge m\ (|s_{n+m} - r_{n+m}| > 2^{-k}).$$

It is easy to próve:

$$\langle r_n\rangle_n < \langle s_n\rangle_n \wedge \langle r_n\rangle_n \sim \langle r'_n\rangle_n \wedge \langle s_n\rangle_n \sim \langle s'_n\rangle_n \rightarrow \langle r'_n\rangle_n < \langle s'_n\rangle_n ,$$

$$\langle r_n\rangle_n \# \langle s_n\rangle_n \wedge \langle r_n\rangle_n \sim \langle r'_n\rangle_n \wedge \langle s_n\rangle_n \sim \langle s'_n\rangle_n \rightarrow \langle r'_n\rangle_n \# \langle s'_n\rangle_n .$$

This justifies the following definition:

6.3.2. Definition.

$$x < y \equiv_D \bigvee \langle r_n\rangle_n \in x \bigvee \langle s_n\rangle_n \in y\ (\langle r_n\rangle_n < \langle s_n\rangle_n),$$

$$x \# y \equiv_D \bigvee \langle r_n\rangle_n \in x \bigvee \langle s_n\rangle_n \in y\ (\langle r_n\rangle_n \# \langle s_n\rangle_n).$$

$\#$ is called the apartness relation for real numbers and represents a kind of positive analogue of the negatively defined inequality $\neq$. It remains to be shown that $\#$ satisfies the properties for an apartness relation as given in § 4.

6.3.3. Theorem. $\#$ is an apartness relation, and

$$x \# y \rightarrow x + z \# y + z;\ z \# 0 \wedge x \# y \rightarrow x.z \# y.z.$$

Proof. [H 1966], 2.2.3, 2.2.5. The proofs of $x = y \rightarrow \neg x \# y$, $x \# y \rightarrow y \# x$, $x \# y \rightarrow x \# z \vee z \# y$ present little difficulties; we shall prove $\neg x \# y \rightarrow x = y$. Let $\langle r_n\rangle_n \in x$, $\langle s_n\rangle_n \in y$, and suppose $(x \# y)$.

Choose n, k such that

$$\bigwedge m\ (|r_n - r_{n+m}| < 2^{-k-2}),\ \bigwedge m\ (|s_n - s_{n+m}| < 2^{-k-2})$$

and suppose $|r_n - s_n| \geq 2^{-k}$.

Then for all m $|s_{n+m} - r_{n+m}| \geq 2^{-k-1}$, i.e. $x \# y$.

This contradicts our assumptions, hence $|r_n - s_n| < 2^{-k}$, and thus $\bigwedge m |r_{n+m} - s_{n+m}| < 2^{-k+1}$. This proves $\langle r_n\rangle_n \sim \langle s_n\rangle_n$, so $x = y$.

The proofs of $x \# y \rightarrow x + z \# y + z$, $z \# 0 \wedge x \# y \rightarrow x.z \# y.z$ are straightforward.

6.3.4. We readily verify that $x \# y \leftrightarrow x < y \vee x > y$.

6.3.5. x^{-1} is clearly defined if $x \# 0$.

In intuitionistic formal systems partially defined operations present a special difficulty; classically we can always make an operation everywhere defined, since the "excluded third" is available. Intuitionistically this does not work; one has to introduce variables for elements belonging to a domain of definition which is undecidable relative to the basic domain of our theory. In the case of x^{-1}, we need in a formal theory special variables for the domain of reals $\# 0$.

§ 7. Ordering relations and order on the real line

7.1

$x < y$ has been defined for real numbers in the previous section.

7.1.1. Definition. $x > y \equiv_D y < x$; $x \nmid y \equiv_D \neg x > y$, $x \nmid y \equiv_D \neg x < y$.

Now we may prove

7.1.2. Theorem.

$$
\begin{array}{ll}
\text{(I)} & x \# y \to x < y \vee x > y, \\
\text{(II)} & x < y \to x \# y, \\
\text{(III)} & x \nmid y \wedge y \nmid x \to x = y \\
\text{(IV)} & x < y \to x < z \vee z < y, \\
\text{(V)} & x < y \wedge y < z \to x < z, \\
\text{(VI)} & x \nmid y \wedge y > z \to x > z, \\
\text{(VII)} & x > y \wedge y \nmid z \to x > z, \\
\text{(VIII)} & x \nmid y \wedge y \nmid z \to x \nmid z.
\end{array}
$$

Proofs. Most of the assertions are easily proved; see [H 1966]2.2.6.

We prove VI as an example. Let $x \nmid y$, $y > z$. There exist k, n such that (supposing $\langle r_n\rangle_n \in x$, $\langle r'_n\rangle_n \in y$, $\langle r''_n\rangle_n \in z$)

$$\bigwedge i (r'_{n+i} - r''_{n+i} > 2^{-k})$$

and a number $m_0 > n$ such that

$$\bigwedge i > m_0 \bigwedge j \, (|r_i - r_{i+j}| < 2^{-k-2}),$$

$$\bigwedge i > m_0 \bigwedge j \, (|r'_i - r'_{i+j}| < 2^{-k-2}).$$

Suppose $i > m_0$, $r_i - r''_i < 2^{-k-2}$. Then $r'_i > r''_i + 2^{-k} > r_i + 3.2^{-k-2}$, and

$$r'_{i+j} > r'_i - 2^{-k-2} > r_i + 2.2^{-k-2} > r_{i+j} + 2^{-k-2}$$

for every j, hence $y > x$. This contradicts our suppositions, hence $r_i - r''_i \geq 2^{-k-2}$; so $r_i - r''_i > 2^{-k-3}$ for every $i > m_0$, hence $x > z$.

7.1.3. Definition. Let $\langle r_n\rangle_n \in x$, $\langle s_n\rangle_n \in y$. $|x|$ is the equivalence class of $\langle |r_n| \rangle_n$, $\max(x,y)$ is the equivalence class of $\langle \max(r_n, s_n)\rangle_n$. Likewise for $\min(x,y)$ ($= \inf(x,y)$).

7.1.4. Theorem. [H 1966] 2.2.7, 2.2.8.

(I) $\max(x,y) + \min(x,y) = x + y$.
(II) $|x| + |y| \nless |x + y|$, $|x - y| \nless \|x| - |y\|$.
(III) $|x| \cdot |y| = |xy|$, $|-x| = |x|$.

7.1.5. Definition. For real numbers x, y:

$[x,y] = \{z : \neg(z > x \wedge z > y) \wedge \neg(z < x \wedge z < y)\}$.

The definition is given in this form because we do not always know whether $x \nless y$ or $x \ngtr y$. By this definition, $[x,y]$ is always inhabited.

7.1.6. Theorem. [H 1966] 3.3.2.

(I) $[x,y] = [\min(x,y), \max(x,y)]$
(II) $x \ngtr y \to [x,y] = \{z : y \nless z \nless x\}$.
(III) $[x,y] = \{z : \max(x,y) \nless z \nless \min(x,y)\}$.

7.2

7.2.1. Definition. Let X be a species, $<$ a binary predicate (relation) defined on a subspecies of X^2, $>$, $\nless$, $\ngtr$ are defined relative to $<$ as in 7.1.1. Consider the following properties (a) - (f):

(a) $x < y \to x \ngtr y \wedge x \neq y$
(b) $x = y \wedge y < z \to x < z$
(c) $x < y \wedge y = z \to x < z$
(d) $x < y \wedge y < z \to x < z$
(e_1) $x = y \vee x < y \vee x > y$
(e_2) $x \nless y \wedge x \ngtr y \to x = y$
(e_3) $x \nless y \wedge x \neq y \to x > y$
(f) $x < y \to \bigwedge z\,(x < z \vee z < y)$.

When $<$ satisfies (a) - (d), $<$ is called a partial ordering relation. $<$ is called an order relation if (a) - (d), (e_1) are satisfied. A relation $<$ which satisfies (a) - (d), (e_2), (f) is called a pseudo-ordering relation. When $<$ satisfies (a) - (d), (e_1), (e_2) it is called a virtual ordering relation.

7.2.2. Remark. If a relation defined on a species is automatically understood as being extensional, then (b), (c) may be omitted everywhere in 7.2.1.

7.2.3. Theorem. Let X be a species with pseudo-ordering $<$. Then the relation $\dot{<}$ defined by $\bigwedge x \in X \bigwedge y \in X (x \dot{<} y \equiv_D x \ngtr y \wedge x \neq y)$ is a virtual ordering ([H 1966] 7.3).

$<$ for reals is a pseudo-order, hence the corresponding $\dot{<}$ is a virtual order. The usual relation $<$ for rationals is an ordering.

The following definition and theorems, due to Brouwer, are interesting because of their formulation in terms of deducibility.

7.2.4. Definition. Let $<'$ be a partial order on a species X, and let Y be the species of valid assertions of the form $x = y$ or $x <' y$ with $x,y \in X$. $<'$ is called unextensible if

(a) $x,y \in X, \neg x <' y$ not deducible from Y by means of (a) - (d) of 7.2.1., then $x <' y \in Y$.

(b) $x,y \in X$, $x \neq y$ not deducible from Y by means of (a) - (d) of 7.2.1., then $x = y \in Y$.

7.2.5. Theorem. ([Br 1927], [H 1966] 7.3.1).
Every virtual ordering relation is unextensible.
Proof. Let X be a species with virtual order $\dot{<}$. Let Y be as in 7.2.4, and suppose $x \dot{>} y \in Y$. Then $x < y$ according to 7.2.1 (a), so if $\neg x \dot{<} y$ cannot be deduced from Y by (a) - (d), then $\neg(x \dot{>} y \in Y)$, and likewise $\neg(x = y \in Y)$, hence by (e_3) $x \dot{<} y \in Y$. Likewise for $=$.

The same kind of reasoning has been used extensively in the proof of
7.2.6. Theorem ([Br 1927]). Every unextensible partial ordering relation is a virtual ordering relation.

7.2.7. Remark. Definition 7.2.4. and the proofs of 7.2.5, 7.2.6. may be rephrased in terms of generalized inductive definitions (§ 15); we shall not execute this here. Presumably, the resulting reformulation represents an application of g.i.d'.s which is weaker than the introduction of K (§ 9) or the introduction of WO (§ 14).

§ 8. Constructive or lawlike analysis

8.1.

In this section we suppose all sequences to be lawlike, although it will turn out that the results obtained in 8.1. - 8.3. are also valid in analysis based on choice sequences (§ 10, 11).

A constructive or lawlike sequence of reals $\langle x_n \rangle_n$ is always supposed to be given by a lawlike sequence $\langle\langle r_{m,n}\rangle_m\rangle_n$, $\langle r_{m,n}\rangle_m \in x_n$ for every n.

8.1.1. Definition. $\langle x_n \rangle_n$ is said to be a Cauchy-sequence if

$$\bigwedge k \bigvee n \bigwedge m (|x_{n+m} - x_n| < 2^{-k}) .$$

$\langle x_n \rangle_n$ is said to be convergent (with limit x) if

$$\bigwedge k \bigvee n \bigwedge m (|x - x_{n+m}| < 2^{-k}).$$

Remark. If $\langle x_n \rangle_n$ is a Cauchy-sequence, then it is clear that also

$$\bigwedge k \bigvee k' \bigwedge n \bigwedge m (n,m \geq k' \rightarrow |x_n - x_m| < 2^{-k}).$$

8.1.2. Theorem. Every Cauchy-sequence $\langle x_n \rangle_n$ is convergent.

Proof. Since $\langle x_n \rangle_n$ is a Cauchy-sequence, there is a lawlike sequence of rationals $\langle\langle r_{m,n}\rangle_m\rangle_n$, $\langle r_{m,n}\rangle_m \in x_n$ for every n, such that

$$\bigwedge k \bigwedge n \bigvee m \bigwedge i \quad (|r_{m+i,n} - r_{m,n}| < 2^{-k}),$$

hence

$$\bigvee a \bigwedge n \bigwedge k \bigwedge m (|r_{a(n,k)+ m,n} - r_{a(n,k),n}| < 2^{-k}).$$

We put

$$s_n = r_{a(n,n),n}$$

$\langle s_n \rangle_n$ is a lawlike real number generator. This is seen as follows. Let $\bigwedge m (|x_{n+m} - x_n| < 2^{-k})$, $n \geq k$, and remark that $|x_m - s_m| \nleq 2^{-m}$, since

$\langle r_{a(m,m)+i,\, m} \rangle_i \in x_m$ and $|x_m - r_{a(m,m),m}| \nleq 2^{-m}$.

Then

$$|s_{n+1} - s_n| \nleq |s_{n+1} - x_{n+1}| + |x_{n+1} - x_n| + |s_n - x_n|$$
$$< 2^{-n-1} + 2^{-k} + 2^{-n} < 3.2^{-k} .$$

Hence there exists a lawlike real number x such that $\langle s_n \rangle_n \in x$.

$$|x - x_{n+1}| \leq |x - s_{n+1}| + |s_{n+1} - x_{n+1}| \leq 3.2^{-k} + 2^{-n-1} < 4.2^{-k} \quad \text{for } n \geq k.$$

8.2.

8.2.1. Definition. A real-valued function f is be supposed to be given by a lawlike functional Ψ such that for real number generators u, v:

$$u \sim v \to \Psi(u) \sim \Psi(v).$$

8.2.2. Definition. A real-valued function f is said to be uniformly continuous if

$$\wedge k \vee m \wedge x \wedge y \, (|x - y| < 2^{-m} \to |fx - fy| < 2^{-k}).$$

f is said to be continuous if

$$\wedge k \wedge x \vee m \wedge y \, (|x - y| < 2^{-k} \to |fx - fy| < 2^{-m}) .$$

It is evident how to formulate continuity and uniform continuity for an interval.

8.2.3. Definition. A real number x is said to be a lowest upper bound (l.u.b.) of a species X of real numbers, if

$$\wedge y \in X (y \leq x \wedge \wedge k \vee y \in X (y > x - 2^{-k})).$$

Clearly, a l.u.b. of a species X, if it exists, is uniquely determined. Not every species of reals can be proved to have an l.u.b.

8.2.4. Definition. x is said to be a l.u.b. for f on $[0, 1]$, iff x is a l.u.b. for $\{fy : y \in [0, 1]\}$.

8.2.5. Theorem. If f is a uniformly continuous function on $[0, 1]$, then f possesses a l.u.b. on $[0, 1]$.

Proof. Consider for every k and every n such that $0 \leq n \leq 2^k$ the number $x_{n,k} = n.2^{-k}$, and put $X_k = \{fx_{n,k} : 0 \leq n \leq 2^k\}$, $\sup X_k = x_k$.
(sup X_k is clearly defined, for if $\{y_0, \ldots, y_n\}$ is a finite species of reals, and $\langle r_n^i \rangle_n \in y_i$ for $0 \leq i \leq n$, then $\langle \sup \{r_m^0, r_m^1, \ldots, r_m^n\} \rangle_m$ is a real number generator for $\sup \{y_0, \ldots, y_n\}$).

We shall prove $\langle x_k \rangle_k$ to be a Cauchy-sequence. Let $a \in (N)N$ be a funtion such that

$$\bigwedge n \bigwedge x \bigwedge y(|x-y| \ngtr 2^{-an} \rightarrow |fx-fy| < 2^{-n}).$$

Then

$$\bigwedge n \bigwedge i(x_{an+i} \nless x_{an}).$$

Let $(x_{an+i} - fx_{an+i,k}) < 2^{-n}$, and let

$$|x_{an+i,k} - x_{an,j}| \ngtr 2^{-an}.$$

Then $|fx_{an+i,k} - fx_{an,j}| < 2^{-n}$; $x_{an} - fx_{an,j} \nless 0$. Therefore

$x_{an+i} - x_{an} = (x_{an+i} - fx_{an+i,k}) + (fx_{an+i,k} - fx_{an,j}) + (fx_{an,j} - x_{an})$

$< 2^{-n} + 2^{-n} = 2^{-n+1}$.

Hence $|x_{an+i} - x_{an}| < 2^{-n+1}$.

By the previous theorem $\langle x_n \rangle_n$ converges to a limit x. This limit x can easily proved to be an l.u.b. for f by manipulating inequalities.

For suppose $fy - x > 2^{-n}$, and let $|y - k.\, 2^{-an}| < 2^{-an}$, then $|fy - f(k.2^{-an})| < 2^{-n}$, so $fy - f(k.2^{-an}) < 2^{-n}$, and since $x \nless x_n \nless f(k.2^{-an})$, it follows that $fy - x < 2^{-n}$; contradiction. Hence $\bigwedge y(fy \ngtr x)$.

Also, since $x - x_{a(n+2)} \ngtr 2^{-n-1}$, and for some k $x_{a(n+2)} - f(x_{a(n+2),k}) < 2^{-n-1}$, it follows that $x - f(x_{a(n+2),k}) < 2^{-n}$.

8.3.

8.3.1. Definition. Let f be a real-valued function on $[0, 1]$, with l.u.b. y. f assumes its l.u.b. (possesses a maximum) if for some real number x, $fx = y$.

8.3.2. We cannot assert that every function uniformly continuous on $[0, 1]$ possesses a maximum. We illustrate this with a "weak counterexample". Let a, b be a lawlike sequences such that

$$\neg(\bigvee n(an = 0) \wedge \bigvee n(bn = 0)).$$

Then we define a real number generator $\langle r_n \rangle_n$ as follows.

$$\neg \bigvee m \leq n\ (am = 0 \vee bm = 0) \to r_n = 0$$
$$am = 0 \wedge \bigwedge m' < m(am' \neq 0) \wedge m \leq n \to r_n = 2^{-m}$$
$$bm = 0 \wedge \bigwedge m' < m(bm' \neq 0) \wedge m \leq n \to r_n = -2^{-m}.$$

Let $\langle r_n \rangle_n \in x_0$. If $\bigvee n(an = 0)$, then $x_0 > 0$; if $\bigvee n(bn = 0)$, then $x_0 < 0$, and if $\neg\bigvee n(an = 0 \vee bn = 0)$, then $x_0 = 0$.

Let us consider the function f given by

$$f(x) = 1 + x_0 . x$$

and suppose that f assumes its l.u.b., say in x_1.
Then $0 < x_1 \vee x_1 < 1$.

If $0 < x_1$, then $\neg\bigvee n(bn = 0)$; if $x_1 < 1$, then $\neg\bigvee n(an = 0)$. So the assumption that every uniformly continuous function assumes its l.u.b. implies

$$\bigwedge a \bigwedge b(\neg(\bigvee n(an = 0) \wedge \bigvee n(bn = 0)) \to (\neg\bigvee n(an = 0) \vee \neg\bigvee n(bn = 0))).$$

Hence also

$$\bigwedge a \bigwedge b((\bigwedge n \neg(\bigvee m\ a\{n, m\} = 0 \wedge \bigvee m\ b\{n, m\} = 0) \to$$
$$\bigwedge n \bigvee n'((\bigvee m\ a\{n, m\} = 0 \to n' = 0) \wedge (\bigvee m\ b\{n, m\} = 0 \to n' \neq 0))),$$

and therefore (5.3,(6))

$$\bigwedge a \bigwedge b((\bigwedge n \neg(\bigvee m\ a\{n, m\} = 0 \wedge \bigvee m\ b\{n, m\} = 0) \to$$
$$\bigvee c \bigwedge n((\bigvee m\ a\{n, m\} = 0 \to cn = 0) \wedge (\bigvee m\ b\{n, m\} = 0 \to cn \neq 0))).$$

Interpreting lawlike functions as recursive functions, we obtain: "Every pair of disjoint r.e. sets can be separated by a recursive set".

8.4.

In intuitionistic mathematics, most classical notions split up into many inequivalent intuitionistic ones (which are provably equivalent when we use classical logic). The splitting-up reflects the differences in constructive content between various classically equivalent definitions. Often there are two interesting cases only: the strongest possible notion with maximal constructive content, and a stable notion (i.e. a property which is equivalent to its double negation). Stable notions may be interesting from a formal point of view.

A simple example was provided by the notions $\#$ and $\neq$ for reals (section 6); $\#$ is the strongest notion, but $\neq$ is stable. (For the relation between $\#$ and $\neq$, see also § 16.6.)

Another example is provided by convergence of sequences of reals: in 8.1.1. the positive notion is defined; a weaker but stable notion is:

8.4.1. <u>Definition</u>. A sequence of reals $\langle x_n \rangle_n$ is said to be <u>negatively</u> convergent (with limit x) if

$$\bigwedge n \neg\neg \bigvee m \bigwedge k(|x - x_{m+k}| < 2^{-n}).$$

The stability of the notion is an immediate consequence of the logical rules $\neg\neg\bigwedge x\, Ax \to \bigwedge x \neg\neg Ax$, $\neg A \leftrightarrow \neg\neg\neg A$.

For some more examples, see e.g. [H 1966], 7.3.2, [R 1960], [R 1963].

8.5.

The difficulties inherent to the general interpretation of e.g. implication often can be avoided by making explicit the extra information required to exist by the premiss of the implication. This is illustrated by theorem of the form

$$\bigwedge f(UC(f) \to A(f)) \tag{1}$$

f a variable for arbitrary real-valued functions on the species of lawlike reals, $UC(f) \equiv_D f$ is uniformly continuous. The validity of $UC(f)$ requires the existence of a modulus of continuity a such that $\bigwedge x \bigwedge y \bigwedge n(|x-y| < 2^{-an} \to |fx-fy| < 2^{-n})$,

and the existence of a sequence $b_0, b_1, b_2, \ldots, b_m$ defined on all rationals $k.2^{-am}$, and taking rational values such that

$|f(k.2^{-am}) - b_m(k.2^{-am})| < 2^{-n}$. This information may be coded into one lawlike $c \in (N)N$, and the coding can be arranged in such a way that conversely every c defines a uniformly continuous f_c (i.e. $\bigwedge c\, UC(f_c)$ and $\bigwedge f(UC(f) \to \bigvee c(f = f_c))$). Now instead of (1) we may prove

$$\bigwedge a(A(f_a))$$

which intuitively expresses the same theorem as (1).

Compare also Kreisel's remarks in [Kr 1968A], appendix.

§ 9. Lawless sequences of natural numbers

9.1.

The material presented in this section is based on [Kr 1968], [Kr 1958]. The behaviour of lawless sequences is radically different from that of lawlike sequences. In this section we shall discuss some important properties of lawless sequences of natural numbers. Throughout this section α, β, γ denote lawless sequences.

To begin with, we introduce a standard enumeration of all finite sequences of natural numbers, including the empty sequence.

Let $\lambda x.\lambda y.\ \{x,y\}$ denote a pairing function with inverses j_1, j_2, such that $\bigwedge z \bigvee x \bigvee y(\{x,y\} = z)$; $\bigwedge x\{j_1 x, j_2 x\} = x$, $\{x,y\} = \{x',y'\} \rightarrow x = x' \wedge y = y'$.
We define inductively enumerations from N^k onto N:

$$\nu_0(x_0) = x_0$$
$$\nu_1(x_0, x_1) = \{x_0, x_1\}$$
$$\nu_u(x_0,\ldots, x_u) = \{x_0, \nu_{u-1}(x_1,\ldots, x_u)\}$$

An enumeration for all finite sequences is obtained by assigning zero to the empty sequence, and assigning

$$\langle x_0,\ldots, x_u\rangle = \{u, \nu_u(x_0,\ldots, x_u)\} + 1$$

to the sequence $x_0,\ldots x_u$. The empty sequence is also written as $\langle\ \rangle$.

lth (n) indicates the length of the sequence with number n. The concatenation function $*$ is introduced by

$$\langle x_0,\ldots, x_u\rangle * \langle x_{u+1},\ldots, x_{u+v}\rangle = \langle x_0,\ldots, x_{u+v}\rangle.$$

We may define $*$ also for concatenation of finite sequences with elements of (N)N:

$$\langle x_0,\ldots, x_{u-1}\rangle * \chi = \chi' \leftrightarrow$$
$$\bigwedge v((v < u \rightarrow \chi' v = x_v) \wedge (v \geq u \rightarrow \chi' v = \chi(v \cdot u))).$$

If y is an initial segment of $\chi (\chi \in (N)N)$, we write $\chi \in y$ or formally

$$\chi \in y \leftrightarrow \bigvee \chi'(\chi = y * \chi')$$

$\bar{\chi}0 = \langle\rangle$, $\bar{\chi}x = \langle \chi 0,\ldots, \chi(x-1)\rangle$ for $x > 0$.

In formulae we shall often write $\hat{x}$ for $\langle x\rangle$, to save space.

Throughout this section, X, Y, Z denote predicates for which all non-lawlike arguments are supposed to be exhibited when they occur in formulae.

9.2.

As a first principle for lawless sequences we state that we can find a lawless sequence with arbitrary prescribed initial segment:

LS1 $\bigwedge x \bigvee \alpha (\alpha \in x)$.

As a consequence, we can find infinitely many different lawless sequences with a prescribed initial segment n, since

$$\bigwedge x \bigwedge \alpha (\alpha \in n * \langle x \rangle \rightarrow \alpha \in n).$$

9.3.

If $\equiv$ is used to denote "identity by definition" (or "intensional equality") i.e. if $\alpha \equiv \beta$ expresses that α, β are to denote the same object (thought-object), then

LS2 $\alpha \equiv \beta \vee \alpha \not\equiv \beta$.

For either we start thinking of α, β as the same object, or we do not.
We also have

$$\alpha \equiv \beta \leftrightarrow \bigwedge x(\alpha x = \beta x) \qquad (1)$$

(or abbreviated: $\alpha \equiv \beta \leftrightarrow \alpha = \beta$).

For if $\alpha \not\equiv \beta$, it is absurd (contradictory) that we could ever prove all values of α, β to be equal, since at any moment (stage) we only know initial segments of α, β. So if we have a proof of $\bigwedge x(\alpha x = \beta x)$, this can be on account of the fact that $\alpha \equiv \beta$ only. The converse implication from the left to the right is evident.

9.4.

For any predicate X:

$$X\alpha \leftrightarrow \bigvee y(\alpha \in y \wedge \bigwedge \beta \in y \; X\beta). \qquad (2)$$

For what is given about a certain individual lawless sequence at a certain stage consists of 1°) its "individuality" or "identity" (expressed by LS2) and 2°) a finite initial segment of values. Hence if we have a property $X\alpha$, α being the only non-lawlike parameter of X, reference to individuality (by setting α equal to another non-lawlike parameter) is excluded, and hence a proof of $X\alpha$ cannot depend on such a reference; therefore $X\alpha$ must be assertable on account of an initial segment only. Thus $X\beta$ must hold for all β with the same initial segment.

More generally, we have for any predicate X:

$$\text{LS3}\quad \begin{cases} \neq(\alpha,\alpha_0,\ldots,\alpha_p) \wedge X(\alpha,\alpha_0,\ldots,\alpha_p) \to \\ \to \bigvee n(\alpha \in n \wedge \bigwedge \beta \in n(\neq(\beta,\alpha_0,\ldots,\alpha_p) \to X(\beta,\alpha_1,\ldots,\alpha_p))) \end{cases}$$

where $\neq(\alpha,\alpha_0,\ldots,\alpha_p)$ is an abbreviation for

$$\alpha \neq \alpha_0 \wedge \alpha \neq \alpha_1 \wedge \ldots \wedge \alpha \neq \alpha_p.$$

The clause $\neq(\alpha,\alpha_0,\ldots,\alpha_p)$ in LS3 serves to exclude reference to "individuality". That this is essential is revealed by the following example. Take for $X(\alpha,\alpha_0)$: $\alpha=\alpha_0$. Then the assertion

$$X(\alpha,\alpha_0) \to \bigvee n(\alpha \in n \wedge \bigwedge \beta \in n\ (\beta = \alpha_0))$$

is evidently false.

(1) and (2) are derivable with the help of LS3. For (2) this is immediate, since it is a special case. For (1) we argue as follows.

$$\alpha = \beta \wedge \alpha \neq \beta \to \bigvee n(\alpha \in n \wedge \bigwedge \alpha' \in n(\alpha' \neq \beta \to \alpha' = \beta))$$

is a consequence of LS3. The conclusion is evidently false, hence $\alpha = \beta \to \neg\alpha \neq \beta$, and so by LS2 we obtain (1).

9.5.

Let us use "LS" to denote the lawless sequences of (N)N, and let Γ, Δ be used for lawlike functionals from LS into a species of lawlike objects (natural numbers, constructive functions, species or relations).
It is a consequence of (2) that such functionals are continuous:

$$\Gamma\alpha = x \leftrightarrow \bigvee n(\alpha \in n \wedge \bigwedge \beta \in n\ (\Gamma\beta = x)) \qquad (3)$$

(Likewise for Γ's of other type).
If we combine (3) with the selection principle:

$$\bigwedge \alpha \bigvee x\ X(\alpha,x) \to \bigvee \Gamma \bigwedge \alpha X(\alpha,\Gamma\alpha).$$

or directly from (2), we obtain a weak form of continuity:

$$\bigwedge \alpha \bigvee x\ X(\alpha,x) \to \bigwedge \alpha \bigvee x \bigvee y \bigwedge \beta(\bar{\alpha}x = \bar{\beta}x \to X(\beta,y)). \qquad (4)$$

We obtain stronger forms of continuity principle if we make stronger assumptions about the class (LS)N (functionals from LS into the natural numbers). For example, it is reasonable to assume that we actually know, given α, if $\Gamma\alpha$ can be computed from $\bar{\alpha}x$, or not.

Essentially stronger, but also plausible, is the assumption that these initial segments $\bar{\alpha}x$ needed to compute $\Gamma\alpha$ can be taken from an a priori given decidable species (to be obtained from a proof of $\bigwedge\alpha \bigvee x\ X(\alpha,x)$). Formally:

$$\bigwedge\Gamma \in (LS)N \bigvee a(\bigwedge\alpha \bigvee x\ a\bar{\alpha}x \neq 0 \wedge \bigwedge n(an \neq 0 \rightarrow \bigvee x \bigwedge\alpha \in n(\Gamma\alpha = x)) \quad (5)$$

and likewise for functionals Γ of other types.

It is easy to verify that we may without loss of generality suppose the a in (5) to satisfy the further requirement:

$$\bigwedge n \bigwedge m(an \neq 0 \rightarrow m = 0 \vee a(n * m) = 0).$$

A further assumption about the lawlike elements of (LS)N (essentially an assumption about the constructive functions a occurring in (5)) is discussed extensively in the next two subsections.

9.6.

We introduce an inductively defined class K of constructive functions of (N)N (which may be interpreted intuitively as neighbourhood functions on Baire space).

Think of a class P of constructive functions which satisfies two closure conditions:

1°) P contains all non-zero constant functions:
$\bigvee x(a = \lambda n.\ x+1) \rightarrow a \in P$

2°) If $a0 = 0$, and for every y $\lambda n.a(\hat{y} * n) \in P$, then $a \in P$, or in a formula:
$a0 = 0 \wedge \bigwedge y(\lambda n.a(\hat{y} * n) \in P \rightarrow a \in P)$.

If we write

$$A_K(P,a) \equiv_D \bigvee x(a = \lambda n.x+1) \vee (\bigwedge y(\lambda n.a(\hat{y} * n) \in P) \wedge a0 = 0)$$

then the closure conditions for P may be expressed by

$$A_K(P,a) \rightarrow Pa . \quad (6)$$

The minimal class K satisfying (6) is exactly the class such that $a \in K$ (or Ka) is proved using (6) only, or in other words, the proof conditions for $a \in K$ are: $a \in K$ is proved using 1° and 2° only. The minimality of K is formally expressed by a schema (P a predicate letter)

$$\bigwedge a[A_K(P,a) \to Pa] \to [K \subseteq P] \ . \qquad (7)$$

Let us look at the structure of K also from a slightly different angle.

A natural "direct" proof of $a \in K$ may be visualized as a (in general infinite) well-founded tree with the topmost node corresponding to the conclusion Ka. Terminal nodes correspond to inferences on account to closure condition 1°.

Passing from a row of immediate descendants of a node ν to ν itself corresponds to an inference on account of (2°). The elements of K occuring in such a proof of $a \in K$ are all of the form $\lambda n.\ a(m * n)$.

In this way a itself represents its own "natural" proof of $a \in K$ (somewhat analogous to the situation for natural numbers).

Induction for natural numbers was justified by a step by step parallelling of the construction of n for every natural number n (see § 3). By a similar argument one can justify (7). Suppose $A_K(P,a)$, $a \in K$.

With every inference of type 1°, 2° in the natural proof of $a \in K$ we associate inferences of type 3°, 4° respectively:

$$3^\circ)\quad \bigvee x(b = \lambda n.x+1) \to b \in K \wedge b \in P$$

$$4^\circ)\quad b0 = 0 \wedge \bigwedge y(\lambda n.b(\hat{y} * n) \in P) \wedge \bigwedge y(\lambda n.b(\hat{y} * n) \in K) \to b \in K \cap P.$$

In this way we obtain by replacing in the natural proof of K every inference of type 1°,or 2° by the corresponding inference of type 3° or 4° a proof of $a \in K \wedge a \in Q$. This justifies (7).

The idea of a "natural" or "direct" proof of $a \in K$ is analogous to the cut-free proofs of proof theory.

9.7.

The elements of K may serve to define continuous functionals of types $((N)N)N$ and $((N)N)(N)N$.

We shall, throughout the remainder of this paper, use e, f to denote elements of K.

One easily proves for elements of K:

$$\bigwedge e \bigwedge \chi \bigvee x(e\bar{\chi}x \neq 0) \tag{8}$$

$$\bigwedge e \bigwedge n \bigwedge m\ (en \neq 0 \rightarrow e(n * m) = en)\ . \tag{9}$$

Let us give the proof for (8). Take Pa to be $\bigwedge\chi\bigvee x(a\bar{\chi}x \neq 0)$.
$P(\lambda n.x+1)$ holds. Suppose

$$a0 = 0,\ \bigwedge y\ P(\lambda n.a(\hat{y} * n)).$$

Take any χ, $\chi = \langle\chi 0\rangle * \chi'$ for a suitable χ'. Hence, since $P(\lambda n.a(\langle\chi 0\rangle * n)$, it follows that $\bigvee y\ (a(\langle\chi 0\rangle * \bar{\chi}'y) \neq 0)$.

Therefore $a\ \bar{\chi}(y + 1) \neq 0$, and thus $\bigwedge\chi\bigvee x(a\bar{\chi}x \neq 0)$, so Pa holds. Thus we have proved $\bigwedge a[A_K(P,a) \rightarrow Pa]$.

By (7) this implies (8). Likewise one proves (9).

From (8), (9) it follows that we may unambiguously define functionals Φ_e, Ψ_e by

$$\Phi_e\chi = y \leftrightarrow \bigvee x(e\bar{\chi}x = y+1)$$

$$(\Psi_e\chi)\ x = y \leftrightarrow \bigvee z(e(\hat{x} * \bar{\chi}z) = y+1).$$

Now we put

$$K^* = \{\Phi_e : e \in K\},\ K^{**} = \{\Psi_e : e \in K\}.$$

The functionals of K^*, K^{**} are everywhere defined on $(N)N$ and continuous. As follows from their mode of generation, the functionals of K^* satisfy (5), since we have evidently:

$$\bigwedge\Gamma \in K^* \bigvee e \in K(\bigwedge\alpha \bigvee x\ e(\bar{\alpha}x) \neq 0 \wedge \bigwedge n(en \neq 0 \rightarrow \bigvee x \bigwedge\alpha \in n(\Gamma\alpha = x)).$$

Now a basic assumption on the structure of the continuous functionals one can make is:

LS4 $\quad K^* = (LS)N$

LS4 is the basic assumption underlying Brouwers argument for the bar theorem and the fan theorem (for choice sequences, [Br 1926 A]), and turns out to be quite strong proof **theoretically**.

The plausibility of LS4 lies in the fact that in order to prove $\Gamma \in (LS)N$, we have to prove $\Gamma\alpha$ to be defined for an arbitrary α on account of an initial segment of α only; no "intensional" information about α can be used. So a proof of $\Gamma \in (LS)N$ must start from a given collection of initial segments n such that $\lambda\alpha.\Gamma(n * \alpha)$ is constant, and then proceed inductively to show $\Gamma\alpha$ to be defined for every α; and this leads naturally to a functional of K^*.

Now our strongest form of the continuity principle for lawless sequences becomes:

$$\text{LS5.} \begin{cases} \bigwedge\alpha_0\bigwedge\alpha_1 \ldots \bigwedge\alpha_p(\#(\alpha_0,\ldots,\alpha_p) \to \bigvee x\, X(x,\alpha_0,\ldots,\alpha_p)) \to \\ \bigvee e \in K \bigwedge\alpha_0 \ldots \bigwedge\alpha_p(\#(\alpha_0,\ldots,\alpha_p) \to X(e(\nu_p(\alpha_0,\ldots,\alpha_p)), \alpha_0,\ldots,\alpha_p)), \end{cases}$$

where $\#(\alpha_0,\ldots,\alpha_p)$ is an abbreviation for $\alpha_i \neq \alpha_j$ for all $i \neq j$, $0 \leq i \leq p$, $0 \leq j \leq p$, and where $e(\chi)$ is defined by

$$e(\chi) = x \leftrightarrow \Phi_e(\chi) = x.$$

Likewise

$$e|\chi = \chi' \leftrightarrow \Psi_e(\chi) = \chi'.$$

A special case of LS5 is

$$\bigwedge\alpha \bigvee x\, X(\alpha,x) \to \bigvee e \bigwedge\alpha X(\alpha, e(\alpha))$$

or **equilently**

$$\bigwedge\alpha \bigvee x\, X(\alpha,x) \to \bigvee e \bigwedge n\, (en \neq 0 \to \bigvee x \bigwedge\alpha \in n X(\alpha,x)).$$

The corresponding principle for constructive functions is

$$\bigwedge\alpha \bigvee a\, X(\alpha,a) \to \bigvee e\, \bigwedge n(en \neq 0 \to \bigvee a\, \bigwedge\alpha \in n\, X(\alpha,a)).$$

Likewise for species etc.

9.8.

Bar theorem. The hypothesis LS4 may also be expressed in a completely different way, namely in the form of the "bar theorem" (the name is more or less a historical accident). The bar theorem may be expressed as a schema:

$$\bigwedge\alpha \bigvee x\, X(\bar{\alpha}x) \wedge \bigwedge n(Xn \to Yn) \wedge \bigwedge n \bigwedge m(Xn \to X(n * m)) \wedge$$
$$\bigwedge n(\bigwedge x Y(n * \hat{x}) \to Yn) \to \bigwedge n\, Yn.$$

<u>Sketch of the proof</u>. Without use of the axioms of the theory of lawless sequences one proves by induction over K with respect to e : $\bigwedge e Ze$, with

$$Ze \equiv_D \bigwedge m \{[\bigwedge n(en \neq 0 \to Y(m * n)) \wedge \bigwedge n(\bigwedge y Y(m * n * \hat{y}) \to Y(m * n))] \to Y(m)\}.$$

Then one easily proves

$$\bigwedge n(en \neq 0 \to Yn) \wedge \bigwedge n(\bigwedge y Y(n * \hat{y}) \to Yn) \to \bigwedge m\, Ym \tag{10}$$

using $\bigwedge e Ze$.

Now suppose $\bigwedge \alpha \bigvee x\, X\, \bar{\alpha}x$, $\bigwedge n(Xn \to Yn)$, $\bigwedge n \bigwedge m(Xn \to X(n * m))$, $\bigwedge n(\bigwedge x\, Y(n * \hat{x}) \to Yn)$. From $\bigwedge \alpha \bigvee x\, X(\bar{\alpha}x)$ we conclude to the existence of an e such that $\bigwedge n(en \neq 0 \to \bigwedge \alpha \in n\, X(\bar{\alpha}(en \dot{-} 1))$.

Then we prove by ordinary induction:

$$\bigwedge n(en \neq 0 \to Yn) . \tag{11}$$

This is done by induction on the length of n.
For $\mathrm{lth}(n) \leq en \dot{-} 1$, (11) is immediate. For the remaining cases, our induction hypothesis is:

$$\bigwedge n(\mathrm{lth}(n) = (en \dot{-} 1) + k \wedge en \neq 0 \to Yn)$$

and this is proved for all k by induction with respect to K. From (11) and $\bigwedge m(\bigwedge y\, Y(m * \hat{y}) \to Ym)$ we can infer, using (10), $\bigwedge m\, Ym$.

Essentially, the bar theorem expresses an induction principle for well-founded species of finite sequences: if we take $R = \{n : \bigwedge m(m <^{o} n \to em = 0)\}$ (with $m <^{o} n \equiv_D \bigwedge m'(n = m * m' \wedge m' \neq 0)$), then

$$\bigwedge n(en \neq 0 \wedge Rn \to Y'n) \wedge \bigwedge n(\bigwedge x\, Y'(n * \hat{x}) \to Y'n) \to \bigwedge n(Rn \to Y'n).$$

This is seen by taking in the formulation of the bar theorem $Xn \equiv_D (en \neq 0)$, $Yn \equiv_D (Rn \to Y'n)$.

On the other hand it is worthwhile remarking that if we <u>define</u> K by $a \in K \equiv_D \bigwedge \alpha \bigvee x(a\bar{\alpha}x \neq 0) \wedge \bigwedge n \bigwedge m(an \neq 0 \to a(n * m) = an)$,
then the axioms for K are provable from the bar theorem.

<u>Sketch of the proof</u>: $\bigwedge a[A_K(K,a) \to Ka]$ is proved straightforwardly.

In order to prove $\bigwedge a[A_K(R,a) \to Ra] \to K \subseteq R$ we suppose $\bigwedge a[A_K(R,a) \to Ra]$, Ka and take $Xn \equiv_D an \neq 0$, $Yn \equiv_D R(\lambda m.e(n * m))$; then we apply the bar theorem for X,Y.

9.9.

Functional relationships between lawless sequences are given by relations X such that

$$\bigwedge \alpha \bigvee ! \beta \quad X(\alpha,\beta,\gamma_0,\ldots,\gamma_n).$$

Let us consider the simplest case:

$$\bigwedge \alpha \bigvee ! \beta \quad X(\alpha,\beta,\gamma).$$

By LS3 we obtain

$$X(\alpha,\beta,\gamma) \to \alpha = \beta \vee \gamma \; = \; \beta \vee \{ \neq (\beta,\alpha,\gamma) \wedge$$
$$\bigvee n(\beta \in n \wedge \bigwedge \beta' \in n(\neq(\beta',\alpha,\gamma) \to X(\alpha,\beta',\gamma)))\}.$$

Since

$$\bigvee n(\beta \in n \wedge \bigwedge \beta' \in n(\neq (\beta',\alpha,\gamma) \to X(\alpha,\beta',\gamma))$$

conflicts with $\bigwedge \alpha \bigvee ! \beta \;\; X(\alpha,\beta,\gamma)$, we have

$$\bigwedge \alpha \bigvee ! \beta \quad X(\alpha,\beta,\gamma) \to \bigwedge \alpha(X(\alpha,\alpha,\gamma) \vee X(\alpha,\gamma,\gamma)). \tag{12}$$

As we shall see, this behaviour makes lawless sequences unfit for a theory of reals and real-valued functions (section 11).

The rigid "lawlessness" of lawless sequences implies that LS is closed under the identity-mapping only.

Even a simple transformation Γ given by $\Gamma\alpha = \lambda x.\alpha(x+1)$ does not transform lawless sequences into lawless sequences; $\Gamma\alpha$ is not lawless since it is in a well-determined way, involving infinitely many function values, related to another sequence α, which cannot be true for lawless sequences.
In fact, suppose $\beta = \Gamma\alpha$. From LS3. it follows that

$$\beta = \Gamma\alpha \to \beta \equiv \alpha \vee \bigvee n[\beta \in n \wedge \bigwedge \gamma \in n(\gamma \neq \alpha \to \gamma = \Gamma\alpha)] \; .$$

Since the second member of the disjunction is evidently false, it follows that

$$\beta = \Gamma\alpha \to \beta \equiv \alpha.$$

But in this case, $\alpha = \beta = \lambda x.\alpha 0$, i.e. a constant function. But $\alpha = \lambda x.y$ implies, by (7):

$$\bigvee n[\alpha \in n \wedge \bigwedge \beta \in n(\beta = \lambda x.y)]$$

which is evidently false. Hence for no lawless α, β $\beta = \Gamma\alpha$.
In fact, analogous to (12) one can prove: $\bigwedge \alpha \bigvee ! \beta \; X(\alpha,\beta) \to \bigwedge \alpha \; X(\alpha,\alpha)$.

9.10.

Lawless sequences are a useful tool in dealing with questions of completeness for intuitionistic logic (see e.g. [K 1958]). Specifically, they provide us with simple refutations of certain classically valid, but intuitionistically invalid logical principles. In this respect they are easier to manage than the choice sequences of section 10.

As an example, we give the following theorem:

Theorem.

(I) $\neg \bigwedge\alpha \bigvee x(\alpha x = 0)$

(II) $\bigwedge\alpha \neg\neg\bigvee x(\alpha x = 0)$

(III) $\neg \bigwedge\alpha(\bigvee x(\alpha x = 0) \vee \neg\bigvee x(\alpha x = 0))$

(IV) $\neg(\bigwedge\alpha\neg\neg\bigvee x(\alpha x = 0) \to \neg\neg\bigwedge\alpha\bigvee x(\alpha x = 0))$

(V) $\neg\bigwedge\alpha(\neg\neg\bigvee x(\alpha x = 0) \to \bigvee x(\alpha x = 0))$.

Proof. (I) Suppose $\bigwedge\alpha \bigvee x(\alpha x = 0)$. It follows from LS5 that for some f

$$\bigwedge n(fn \neq 0 \to \bigwedge\alpha \in n(\alpha(fn \dot{-} 1) = 0)$$

and if we determine u, m such that

$$m = \overline{(\lambda y.1)}u, \; fm \neq 0$$

then $\bigwedge\alpha \in m'(\alpha(fm \dot{-} 1) = 0)$ for $m' = (\overline{\lambda y.1})(u + (fm \dot{-} 1))$

which is plainly contradictory.

(II) Suppose $\neg\bigvee x(\alpha x = 0)$. By (7),

$$\neg\bigvee x(\alpha x = 0) \leftrightarrow \bigvee n(\alpha \in n \wedge \bigwedge\beta \in n \neg\bigvee x(\beta x = 0)).$$

This is evidently false, hence $\neg\neg\bigvee x(\alpha x = 0)$.

(III) $\neg\bigvee x(\alpha x = 0)$ is contradictory, so if we suppose $\bigwedge\alpha(\bigvee x(\alpha x = 0) \vee \neg\bigvee x(\alpha x = 0)$ it follows that $\bigwedge\alpha \bigvee x(\alpha x = 0)$, which conflicts with (I). Likewise we obtain (IV), (V) as simple consequences of (I), (II).

This theorem therefore refutes the logical principles $A \vee \neg A$, $\bigwedge x \neg\neg A \to \neg\neg\bigwedge x\, A$, and "Markov's principle" for lawless sequences, by II, IV, V. (Markov's principle in its proper sence only states something for mechanically computable functions).

§ 10. Choice Sequences

10.1.

Choice sequences are intermediate in character between lawless and lawlike sequences.

Lawless sequences have shown themselves to possess a strong "spirit of freedom" that they cannot be linked by a non-trivial relationship involving infinitely many values.

Lawlike sequences produced decent analysis, but there we had no opportunity to exploit the fact that most operations on sequences are continuous. Lawlike analysis made us suspect that every well-defined real-valued function is continuous, but we were unable to prove it.

Choice sequences possess some freedom, but they are "on leash" so to speak; we can restrict their freedom, or even bring them completely to reason and force them to be (extensionally) equal to a lawlike sequence. In analysis based on choice sequences, real-valued functions are continuous.

Brouwer does not spend much words on telling us what exactly a choice sequence is ([Br 1924], footnote 3 on page 245, [Br 1942], [Br 1952] footnote* on page 142). His notion of sequence is approximately described by saying that we choose values and restrictions in the form of a spread-law (see the end of this section) on future choices of values; for a recent discussion of this notion see Myhill's paper [M 1967]; see also [T 1968], page 220.

The concept of a choice sequence as it is discussed here is probably neither the only possible nor the simplest one, but among the alternatives discussed so far, it seems to me to be the most fruitful notion for analysis. The formal system CS discussed in [T 1968] and the system of the monograph of Kleene and Vesley [K, V 1965] may be interpreted by means of this notion.

For lack of a better name, I have baptized the notion of choice sequence discussed here, GC-sequences (from: Generated by Continuous operations). In the sequel, "choice sequence" and "GC-sequence" will be regarded as synonyms.

There is a definite need for a formal development in which the basic notions which occur in the intuitive justification occur explicitly. It may well be that a detailed formal analysis would reveal some hidden assumptions, but I believe nevertheless that the notion of choice sequence as discussed here is a good approximation of what we need. Besides, at present I do not have a more satisfactory alternative available.

10.2.

By associating with the choice sequences of (N)N, to be introduced below, real number generators, we shall obtain a theory of (choice-) real numbers (see § 11).

Working in a theory of real numbers, we require certain closure properties e.g. $x, y \in \mathcal{R} \to x, y \in \mathcal{R}$.

On the other hand, when speaking about an arbitrary individual real number x, it is natural in the course of our argument to require specifications like "x is such that $\bigvee z(x = f(z))$", and later on e.g. "$x = f(z)$; z is such that $g(u) = z$ for some u" (f, g lawlike functions).

This suggests the following notion of choice sequence. (We use $\alpha, \beta, \gamma, \delta$ for choice sequences throughout this section.)

A choice sequence α is obtained by chosing values $x_0, x_1, \ldots$; we may decide at a certain stage to restrict future choices of values by requiring $\alpha = \Gamma_0\alpha_0$, where α_0 is a "fresh", newly introduced choice sequence (that is, a new process for computing values is started), Γ_0 a continuous operation (one may think of a functional of K^{**}); for some time, we compute further values of α by selecting values of α_0; at a later stage, we may again decide to restrict α_0 by selecting a continuous operator Γ_1 and requiring $\alpha_0 = \Gamma_1\alpha_1, \alpha_1$ a newly started choice sequence, and so on.

We now present this description in a more schematic way. We think of a choice sequence α as a sequence of pairs:

$$\alpha \equiv \langle x_0, R_0\rangle, \langle x_1, R_1\rangle, \langle x_2, R_2\rangle, \ldots$$

which are successively generated; the x_i are natural numbers, the R_i are (extensional) conditions on the sequence $\langle x_n\rangle_n$ of numerical values of α. The process of generation of pairs must satisfy some general a priori conditions (numerical values chosen must be consistent with already chosen, an R_{x+1} following an R_x must satisfy a certain relation $\sqsubset(R_{x+1}, R_x)$, the R_x must be taken from a certain class $\mathcal{R}$), but for the rest we know at any stage only an initial segment of α. For our notion of choice sequence, the R_i stipulate relationships to other choice sequences.

Extensional equality of choice sequences $\alpha \equiv \langle\langle x_n, R_n\rangle\rangle_n$ and $\beta \equiv \langle\langle y_n, R'_n\rangle\rangle_n$ may be expressed as

$$\bigwedge n(x_n = y_n) \leftrightarrow \alpha = \beta.$$

For any $\alpha \equiv \langle\langle x_n, R_n\rangle\rangle_n$ we write αn for x_n, R_n^α for R_n.
The conditions R_i are taken from a special class $\mathcal{R}$; $\mathcal{R}$ consists of all conditions of the form:

$$R \equiv \lambda\alpha.(\alpha = \Gamma_0\alpha_0 \wedge \alpha_0 = \Gamma_1\alpha_1 \wedge \ldots \wedge \alpha_{n-1} = \Gamma_n\alpha_n) \tag{1}$$

where the $\Gamma_0, \ldots, \Gamma_n$ are continuous operations on sequences.

Furthermore we require that the R_i used in the description of α satisfy the condition: if R_i has the form (1), then R_{i+1} must have the form

$$\lambda\alpha.(\alpha = \Gamma_0\alpha_0 \wedge \ldots\ \alpha_{n-1} = \Gamma_n\alpha_n \wedge \alpha_n = \Gamma_{n+1}\ \alpha_{n+1} \wedge \ldots \wedge \alpha_{n+p-1} = \Gamma_{n+p}\ \alpha_{n+p}).$$

Now the picture as presented above is too simple, in the sense that at any stage α depends essentially on one other choice sequence only. There is also the possibility of letting α depend on more than one other sequence in a given stage.

Let us put

$$\nu_u(\alpha_0, \ldots, \alpha_u) = \lambda x.\ \nu_u(\alpha_0 x, \ldots, \alpha_u x).$$

Now an example of the more general type of condition is:

$$R \equiv \lambda\alpha.[\alpha = \Gamma_0\alpha_0 \wedge \alpha_0 = \Gamma_0\ \nu_2(\alpha_1,\alpha_2) \wedge \alpha_1 = \Gamma_1\ \nu_2(\alpha_3, \alpha_4) \wedge \alpha_2 = \Gamma_2\ \nu_2(\alpha_5,\alpha_6)]$$

or in general

$$R \equiv \lambda\alpha.[\alpha = \Gamma\nu_m(\alpha_1, \ldots, \alpha_m) \wedge \bigwedge_{i=1}^{m} \alpha_i = \Gamma_i(\alpha_{i1}, \ldots, \alpha_{i\,m_i}) \wedge$$
$$\wedge \bigwedge_{i=1}^{m} \bigwedge_{j=1}^{m_i} \alpha_{ij} = \Gamma_{ij}\ \nu_{m_{ij}}\ (\alpha_{ij1}, \ldots, \alpha_{ij\,m_{ij}}) \wedge \ldots] .$$

But this extension of the class $\mathcal{R}$ does not essentially alter the discussion of the principles valid for choice sequences, given below, hence we restrict the discussion to the simplified version.

Note that if at a given stage $\alpha = \Gamma\nu_2(\alpha,\beta)$, then it is always possible that we decide later on $\alpha = \lambda x.j_1\gamma x$, $\beta = \lambda x.j_2\gamma x$ (or in short $\alpha = j_1\gamma$, $\beta = j_2\gamma$), so then $\alpha = \Gamma'\gamma$.

From the preceding description it will be clear that continuous operations transform choice sequences into choice sequences.

For if $\alpha \equiv \langle\langle \alpha x, R_x^{\alpha} \rangle\rangle_x$, with say $R_0^{\alpha} \equiv \lambda\alpha[\alpha = \Gamma_0\alpha_0]$, $R_1^{\alpha} \equiv \lambda\alpha[\alpha = \Gamma_0\alpha_0 \wedge \alpha_0 = \Gamma_1\alpha_1]$,

and if we start $\beta = \Gamma\alpha$ as $\beta \equiv \langle \beta 0, R_0^{\beta} \rangle$,... with $R_0^{\beta} \equiv \lambda\beta.[\beta = \Gamma\alpha \wedge \alpha = \Gamma\alpha_0]$

then the necessary

information to insure $\beta = \Gamma\alpha$ is completely contained in R_0^{β} (hence in an initial segment of pairs of β), and we are free to extend β (within the general a priori conditions) so β may conceived as a choice sequence too.

It is perhaps more appropriate to describe the process of generation not as the generation of a single choice sequence, but as the generation of a "network" of choice sequences which are interconnected by conditions of the form (1). Universal statements $\bigwedge\alpha A\alpha$ are then to be interpreted as "for all α in all possible networks $A\alpha$" etc.

Remark. On a formal scheme of description as used above it is possible to construct a wide variety of notions of sequence.

10.3.

The principle of intensional continuity. From now on, X, Y, Z are used in this section for extensional predicates with all non-lawlike variables exhibited when they occur in formulae. Extensionality means

$$X(\alpha_0,\ldots,\alpha_i,\ldots) \wedge \alpha_i = \beta \rightarrow X(\alpha_0,\ldots,\alpha_{i-1},\beta,\alpha_{i+1},\ldots).$$

A first principle that is valid for choice sequences is the following. Let X be an extensional predicate. Then

$$X\alpha \leftrightarrow \bigvee\Gamma(\bigvee\beta(\alpha = \Gamma\beta) \wedge \bigwedge\beta\, X(\Gamma\beta)) \,. \qquad (2)$$

The implication from the right to the left is trivial. Suppose that we can assert $X\alpha$. Then this must be possible on account of an initial segment of $\alpha: \langle x_0, R_0\rangle, \ldots, \langle x_n, R_n\rangle$.

Suppose e.g. $R_n\alpha$ to have the form $\alpha = \Gamma'\alpha' \wedge \alpha' = \Gamma''\alpha''$. If at the time we assert $X\alpha$, α'' is as yet completely undetermined, then α'' is entirely arbitrary, hence $\bigwedge\gamma\, X(\Gamma'\Gamma''\gamma)$.

If α'' is determined up to $n = \langle \alpha''0, \ldots, \alpha''x\rangle = \langle v_0, \ldots, v_x\rangle$ say, then α'' may always be conceived as being obtained from α'' by a special continuous operation $\Gamma^{(n)}$ such that

$$\bigwedge x \bigwedge y\{(y \leq x \rightarrow (\Gamma^{(n)}\chi)(y) = v_y) \wedge (y > x \rightarrow (\Gamma^{(n)}\chi)y = \chi y)\}.$$

So α may be conceived as being obtained as $\alpha = \Gamma'\Gamma''\Gamma^{(n)}\alpha'''$, hence $\bigwedge\gamma X(\Gamma'\Gamma''\Gamma^{(n)}\gamma)$.

10.4.
Extensional continuity

The next principle for choice sequences to be discussed is the axiom of extensional continuity (called Brouwer's principle in [K, V 1965]):

$$\bigwedge\alpha \bigvee x\, X(\alpha,x) \to \bigwedge\alpha \bigvee x \bigvee y \bigwedge\beta(\bar{\alpha}x = \bar{\beta}x \to X(\beta,y)) \,. \qquad (3)$$

Suppose $\bigwedge\alpha \bigvee x\, X(\alpha,x)$. A proof of $\bigwedge\alpha \bigvee x\, X(\alpha,x)$ must contain a procedure Ψ which computes a natural number $\Psi\alpha$ for an arbitrary α such that $X(\alpha,\Psi\alpha)$. Thus we have

$$\bigwedge\alpha\, X(\alpha,\Psi\alpha).$$

In general Ψ is an intensional operation, i.e. Ψ makes use of values as well as restrictive conditions introduced during the generation of the choice sequences.

Now we introduce a certain conceptual operator, an abstraction process Abstr which transforms a choice sequence α into another choice sequence Abstr α.

If $\alpha \equiv \langle\langle x_n, R_n\rangle\rangle_n$, then we think of Abstr (α) as a choice sequence

$$\langle x_0, U\rangle, \langle x_1, U\rangle, \ldots$$

where U indicates the universal condition $\lambda\alpha.\ \alpha = \alpha$, and where we forget the mode of generation from α. That is, at any moment we consider all future restrictions possible for α, because we abstract from α by omitting systematically the restrictions. Remark that Abstr (α) is not a lawless sequence, since the "thought object" Abstr (α) is such that we think at any stage future restrictions possible.

Maybe it is more accurate to say that Abstr (α) is not simply a choice sequence, but an object which is indistinguishable from a choice sequence w.r.t. mathematical operations, or yet more precise: each operation defined on choice sequences is automatically defined for objects like Abstr (α) too.

We cannot assert α = Abstr (α), since we must necessarily know the mode of generation of Abstr (α) from α for this. But on the other hand, for any given $x \in N$ it is effectively verifiable that

$$\bar{\alpha}x = \overline{\mathrm{Abstr}\,(\alpha)}\,x \;.$$

Formulating this in yet another way, abstracting from the mode of generating Abstr α means that in applying any kind of mathematical operation on Abstr (α), we cannot make use of α = Abstr (α), but at any stage we can make use of $\bar{\alpha}x = \overline{\text{Abstr}\ (\alpha)}\ x$ for any initial segment $\overline{\text{Abstr}\ (\alpha)}x$ computed at the given stage.

It is useful here to point to one of the specific problems one encounters in an attempt to formalize Abstr .
Although it is permissible to assert

$$\bar{\alpha}t = \overline{\text{Abstr}\ (\alpha)}(t)$$

for any constant term t, we may not assert

$$\bar{\alpha}x = \overline{\text{Abstr}\ (\alpha)}x$$

and then use the rule of generalization to obtain

$$\bigwedge x(\bar{\alpha}x = \overline{\text{Abstr}\ (\alpha)}\ x)$$

since this implies α = Abstr (α).

Now suppose Ψ Abstr $(\alpha) = x$ to be determined from an initial segment $\langle x_0, U\rangle, \ldots, \langle x_n, U\rangle$. Hence for any β beginning with $\langle x_0, U\rangle, \ldots, \langle x_n, U\rangle, \langle\alpha_{n+1}, R_{n+1}\rangle$, $X(\beta,x)$ must hold.

Since $R_{n+1}\beta \wedge R_{n+1}\alpha$ implies $\alpha = \beta$, it follows that $X(\alpha,x)$ must hold i.e.

$$\bigwedge\alpha\ X(\alpha,\Psi(\text{Abstr}\ (\alpha))).$$

Since Ψ(Abstr (α)) depends on initial segments of numerical values only, we conclude for any α to

$$\bigvee x \bigvee y \bigwedge\beta(\bar{\alpha}x = \bar{\beta}x \rightarrow X(\beta,x)),$$

and thus we have justified (3). Expressed otherwise

$$\bigwedge\alpha\ \bigvee x\ X(\alpha,x) \rightarrow \bigvee\Gamma\bigwedge\alpha\ X(\alpha,\Gamma\alpha), \tag{4}$$

Γ being a lawlike continuous operator from choice sequences (the species GC) into N. For the same reasons as in the case of lawless sequences (specifically: since Γ has to be defined on all sequences Abstr (α)) we may introduce here the assumption

$$K^* = (GC)N\ . \tag{5}$$

As in the case of lawless sequences, there is an intermediate possibility of strengthening (3):

$$\bigwedge\Gamma \in (GC)N\ \bigvee a(\bigwedge\alpha\ \bigvee x\ \ a\bar{\alpha}x \neq 0 \wedge \bigwedge n(an \neq 0 \rightarrow \bigvee x\ \bigwedge\alpha \in n(\Gamma\alpha = x))\ . \tag{6}$$

10.5.

Formal consequences of the principle discussed.

We remark that a lawlike Γ of type (GC)GC belongs to K^{**}, since $\bigwedge x(\lambda\alpha.(\Gamma\alpha)x \in K^{*})$, i.e. $\bigwedge x \bigvee e \bigwedge\alpha((\Gamma\alpha)x = e(\alpha))$, hence $\bigvee f \bigwedge x \bigwedge\alpha(e(\alpha) = \lambda n.f(\hat{x} * n)(\alpha))$. Therefore the axiom of intensional continuity combined with (5) reads:

$$X\alpha \to \bigvee e[\bigvee\beta(e|\beta = \alpha) \wedge \bigwedge\gamma\, X(e|\gamma)]. \tag{7}$$

An important consequence of (7) is

$$\bigwedge\alpha[X\alpha \to Y\alpha] \leftrightarrow \bigwedge e[\bigwedge\alpha\, X(e|\alpha) \to \bigwedge\alpha Y(e|\alpha)] . \tag{8}$$

Proof of (8). The implication from the left to the right is immediate. Suppose conversely $\bigwedge e[\bigwedge\alpha\, X(e|\alpha) \to \bigwedge\alpha Y(e|\alpha)]$, $X\alpha$. Then for some f, β $\alpha = f|\beta \wedge \bigwedge\gamma\, X(f|\gamma)$. Hence $\bigwedge\gamma\, Y(f|\gamma)$, and thus $Y\alpha$.

The principle of extensional continuity (3) can be strengthened to the case of an X with parameters:

$$\bigwedge\alpha\bigvee x\, X(\alpha,x,\alpha_0,\dots,) \to \bigwedge\alpha\bigvee x\bigvee y\bigwedge\beta(\bar{\alpha}x = \bar{\beta}x \to X(\beta,y,\alpha_0,\dots)). \tag{9}$$

Proof of (9). Let us take the case with a single extra parameter:

$$\bigwedge\beta[\bigwedge\alpha\bigvee x\, X(\alpha,x,\beta) \to \bigwedge\alpha\bigvee x\bigvee y\bigwedge\alpha'(\bar{\alpha}'y = \bar{\alpha}y \to X(\alpha',x,\beta))].$$

By (8) this is equivalent to

$$\bigwedge e[\bigwedge\beta\bigwedge\alpha\bigvee x\, X(\alpha,x,e|\beta) \to \bigwedge\beta\bigwedge\alpha\bigvee x\bigvee y\bigwedge\alpha'(\bar{\alpha}'y = \bar{\alpha}y \to X(\alpha',x,e|\beta))].$$

Suppose $\bigwedge\beta\bigwedge\alpha\bigvee x\, X(\alpha,x,e|\beta)$.

Since $\bigwedge\gamma\bigvee\alpha\bigvee\beta(j_1\gamma = \alpha \wedge j_2\gamma = \beta)$ and $\bigwedge\alpha\bigwedge\beta\bigvee\gamma(\{\alpha,\beta\} = \gamma)$

(closure of the notion of choice sequence under continuous operations) we have

$$\bigwedge\alpha\bigwedge\beta\bigvee x\, X(\alpha,x,e|\beta) \leftrightarrow \bigwedge\gamma\bigvee x\, X(j_1\gamma,x,e|j_2\gamma).$$

Therefore from our supposition

$$\bigwedge\gamma\bigvee x\, X(j_1\gamma,x,e|j_2\gamma)$$

hence, using (3):

$$\Lambda\gamma\, V x V y \Lambda\gamma' (\overline{j_1\gamma} y = \overline{j_1\gamma'} y \rightarrow X(j_1\gamma', x, e|j_2\gamma'))$$

and thus

$$\Lambda\alpha\Lambda\beta V x V y \Lambda\alpha'\Lambda\beta' (\bar{\alpha} y = \bar{\alpha}' y \rightarrow X(\alpha', x, e|\beta'))$$

so

$$\Lambda\beta'\Lambda\alpha V x V y \Lambda\alpha' (\bar{\alpha} y = \bar{\alpha}' y \rightarrow X(\alpha', y, e|\beta'))$$

and thus we have proved the case with a single extra parameter. Likewise in the case of more parameters; e.g.

$$\Lambda\beta\Lambda\gamma [\Lambda\alpha V x\; X(\alpha, x, \beta, \gamma) \rightarrow \Lambda\alpha V x V y \Lambda\alpha' (\bar{\alpha}' y = \bar{\alpha} y \rightarrow X(\alpha', x, \beta, \gamma))]$$

is reduced to the previous case by first applying

$$\Lambda\beta\Lambda\gamma\; Y(\beta, \gamma) \leftrightarrow \Lambda\delta Y(j_1\delta,\; j_2\delta).$$

Remark. The method used to prove (9) is applicable to many cases where a result without parameters has to be generalized to the corresponding result with extra choice parameters.

We mention especially:

$$\Lambda\alpha V x\; X(\alpha, x, \alpha_0, \dots, \alpha_{u-1}) \rightarrow V e \Lambda\alpha\; X(\alpha, e(\nu_u(\alpha, \alpha_0, \dots, \alpha_{u-1})),\; \alpha_0, \dots, \alpha_{u-1}). \quad (10)$$

The bar theorem for X, Y not containing choice variables is proved in exactly the same manner as in the case of lawless sequences. The bar theorem for arbitrary X, Y is contained from the special case by a reasoning analogous to the argument in the proof of (9).

The derivation of the following formula deserves special attention

$$\Lambda x V y\; X(x, y, \alpha_0, \dots) \rightarrow V\beta\Lambda x\; X(x, \beta x, \alpha_0, \dots)\,. \quad (11)$$

(11) is obtained from the special case (justified in § 5)

$$\Lambda x V y\; X(x, y) \rightarrow V b \Lambda x\; X(x, bx).$$

In order to prove (11), we do not use (8), but (7) instead.

Proof of (11). Let $\Lambda x V y\; X(x, y, \alpha)$. Then there is an f and a γ such that $f|\gamma = \alpha$, $\Lambda\delta\Lambda x V y\; X(x, y, f|\delta)$. We use (3) and obtain $\Lambda x V e \Lambda\delta\; X(x,\; e(\delta), f|\delta)$, and therefore also $\Lambda x \Lambda\delta\; X(x, \lambda n. e'(\hat{x} * n)(\delta),\; f|\delta)$ for some $e' \in K$.

Define $\beta = \lambda x.(\lambda n.e'(\hat{x} * n)(\delta)) = e'|\delta$ and take $\delta = \gamma$, then it follows that $\bigwedge x\, X(x, \beta x, \alpha)$ and this proves (11).

10.6.
Derivation of the fan theorem.

10.6.1. Definition. A lawlike function a is said to represent a spread-law (notation: Spr (a) or $a \in$ Spr) if

(a) $a0 \neq 0$
(b) $\bigwedge n \bigwedge m(a(n * m) \neq 0 \to an \neq 0)$
(c) $\bigwedge n \bigvee x(an \neq 0 \to a(n * \hat{x}) \neq 0)$.

The spread-law is said to be finitary if in addition
(d) $\bigwedge n \bigvee z \bigwedge x(a(n * \hat{x}) \neq 0 \to x \leq z)$.
A finitary spread-law is also called a fan-law.

Intuitively, a spread-law represents a set of nodes $\{n : an \neq 0\}$ of a "tree" of finite sequences of natural numbers, with branches directed downwards; the topmost node corresponds always with the empty sequence. A fan-law corresponds to a finitely branched tree.

10.6.2. Definition. For any $a \in$ Spr, $\chi \in (N)N$:

$$\chi \in a \equiv_D \bigwedge n(a(\bar{\chi}n) \neq 0).$$

10.6.3. Definition. With any spread-law a we may associate a continuous operation Γ_a, represented by an element $e_a \in K$, as follows. Let b be a function such that

$$\bigwedge n \bigwedge x(a(n * \hat{x}) \neq 0 \to a(n * \langle bn \rangle) \neq 0 \wedge bn \leq x).$$

Then there exists a unique c such that

$$\bigwedge n \bigwedge x[(a(n * \hat{x}) \neq 0 \to c\{n, x\} = x) \wedge (a(n * \hat{x}) = 0 \to c\{n, x\} = bn)]$$

and unique d, d_1 such that

$$d0 = 0, \quad d_1 0 = 0$$
$$d_1(n * \hat{x}) = c\{dn, x\}, \; d(n * \hat{x}) = dn * c\{dn, x\}.$$

Now $\Gamma_a\chi = \lambda x.d(\bar{\chi}(x + 1))$. e_a may be defined by

$$e_a\langle z, x_0,\ldots, x_v\rangle = 0 \text{ if } v < z$$
$$e_a\langle z, x_0,\ldots, x_v\rangle = d\langle x_0,\ldots, x_z\rangle + 1 \text{ if } z \leq v.$$

10.6.4. Remark. If α is any choice sequence such that $\alpha \in a$, $a \in \underline{\mathrm{Spr}}$, then $\Gamma_a\alpha = \alpha$. For all α $\Gamma_a\alpha \in a$, hence $\Gamma_a\Gamma_a\alpha = \Gamma_a\alpha$.

10.6.5. Lemma. For any fan-law a:

$$\bigwedge e \bigvee z \bigwedge \alpha(e(\Gamma_a\alpha) \leq z).$$

Proof. In order not to obscure the structure of the proof, we give the proof in a simple case, by taking for a the binary spread law:

$$a\langle x_0,\ldots, x_v\rangle = 1 \leftrightarrow \bigwedge i \leq v(x_i \leq 1), \bigwedge n(an \in \{0,1\}).$$

Γ_a is described by

$$\Gamma_a\chi = \lambda x.(1 \dot{-} |\chi x - 1|).$$

Let us put $h = \lambda x.1 \dot{-} |x - 1|$ (so $hx = 0$ if $x = 0$ or $x > 1$, $hx = 1$ if $x = 1$). Let $\Gamma = \Gamma_a$ in the sequel. We prove by induction over K with respect to e: $\bigwedge e \bigvee z \bigwedge \alpha(e(\Gamma\alpha) \leq z)$.

If $e = \lambda n.x + 1$, then $\bigwedge \alpha(e(\Gamma\alpha) \leq x)$. Suppose now:

$$e0 = 0,\ \bigwedge x \bigvee z \bigwedge \alpha[\lambda n.e(\hat{x} * n)(\Gamma\alpha) \leq x].$$

We have $h\alpha 0 = 0 \vee h\alpha 0 = 1$, and we can find x_0, x_1 such that

$$\bigwedge \alpha(e_0(\Gamma\lambda x.\alpha(x + 1)) \leq x_0)$$
$$\bigwedge \alpha(e_1(\Gamma\lambda x.\alpha(x + 1)) \leq x_1)$$

where $e_0 = \lambda x.e(\hat{0} * x)$, $e_1 = \lambda x.e(\hat{1} * x)$. Therefore

$$\bigwedge \alpha(e_{h\alpha 0}(\Gamma\lambda x.\alpha(x + 1)) \leq \sup(x_0, x_1)$$

hence

$$\bigwedge \alpha(e(\Gamma\alpha) \leq \sup(x_0, x_1))$$

so

$$\bigvee z \bigwedge \alpha(e(\Gamma\alpha) \leq z).$$

In the general case the argument is slightly more complicated; here it is more advantageous to apply induction over K w.r.t. e to $\bigwedge e \bigwedge a$(a is a fan-law $\rightarrow$ $\bigvee z \bigwedge \alpha(e(\Gamma_a \alpha) \leq z))$.
If d is the function as indicated in the definition of Γ_a we use $e(\Gamma_a \alpha) = \lambda n.e(d\langle\alpha 0\rangle * n)(\Gamma_b \lambda x.\alpha(x + 1))$, where b is the fan-law $\lambda n.a(d\langle\alpha 0\rangle * n)$, in the induction step. The details are left to the reader.

10.6.6. Remark. It is easy to see that once we have proved this lemma, we also have for any fan-law a and any Γ such that $\bigwedge \alpha(\Gamma\alpha \in a)$ and $\bigwedge \beta \in a \bigvee \alpha(\Gamma\beta = \alpha)$: $\bigwedge e \bigvee z \bigwedge \alpha(e(\Gamma\alpha) \leq z)$.

10.6.7. Theorem (fan-theorem). Let a be a fan-law. Then

$$\bigwedge \alpha \bigvee x\, X(\Gamma_a \alpha, x, \alpha_0, \ldots) \rightarrow \bigvee y \bigwedge \alpha \bigvee x \bigwedge \beta(\overline{\Gamma_a \alpha}y = \overline{\Gamma_a \beta}y \rightarrow X(\Gamma_a \beta, x, \alpha_0, \ldots)).$$

Proof. Let α be the only non-lawlike variable in X, and suppose $\bigwedge \alpha \bigvee x\, X(\Gamma_a \alpha, x)$. Then we can find an e such that $\bigwedge \alpha\, X(\Gamma_a \alpha, e(\alpha))$, hence $\bigwedge \alpha\, X(\Gamma_a \alpha, e(\Gamma_a \alpha))$. Then there exists an f such that $\bigwedge \alpha(e(f(\alpha)) \neq 0)$. We apply the lemma 10.6.5 and obtain a y such that $\bigwedge \alpha(f(\Gamma_a \alpha) \leq y)$. Hence for any α

$$\bigwedge \beta(\overline{\Gamma_a \alpha}y = \overline{\Gamma_a \beta}y \rightarrow X(\Gamma_a \beta, e(\Gamma_a \alpha))).$$

Extension to the general case with parameters $\alpha_0, \alpha_1, \ldots$ as in the proof of (9).

10.6.8. Remark. For any Γ such that $\bigwedge \alpha(\Gamma\alpha \in a)$, $\bigwedge \beta \in a \bigvee \alpha(\Gamma\beta = \alpha)$, we have $\bigwedge \alpha \bigvee x\, X(\Gamma\alpha, x) \rightarrow \bigvee y \bigwedge \alpha \bigvee x \bigwedge \beta(\overline{\Gamma\alpha}y = \overline{\Gamma\beta}y \rightarrow X(\Gamma\beta, y))$.

As in the theory of lawless sequences, we may also obtain continuity principles corresponding to quantifier combinations $\bigwedge \alpha \bigvee a$, $\bigwedge \alpha \bigvee X^\sigma$ (X^σ a species variable with a specified "signature" σ of arguments).

Thus we obtain a.o.

$$\bigwedge \alpha \bigvee a\, X(\alpha, a, \alpha_0, \ldots) \rightarrow \bigwedge \alpha \bigvee x \bigvee a \bigwedge \beta(\bar{\alpha}x = \bar{\beta}x \rightarrow X(\beta, a, \alpha_0, \ldots)) \tag{12}$$

$$\bigwedge \alpha \bigvee X^\sigma Y(\alpha, X^\sigma, \alpha_0, \ldots) \rightarrow \bigwedge \alpha \bigvee x \bigvee X^\sigma \bigwedge \beta(\bar{\alpha}x = \bar{\beta}x \rightarrow Y(\beta, X^\sigma, \alpha_0, \ldots)) \tag{13}$$

$$\bigwedge \alpha \bigvee a\, X(\alpha, a) \rightarrow \bigvee e \bigwedge n(en \neq 0 \rightarrow \bigvee a \bigwedge \alpha \in n\, X(\alpha, a)) \tag{14}$$

etc. etc.

One would expect for $\Lambda\alpha V\beta$ the following form of continuity:

$$\Lambda\alpha V\beta\, X(\alpha, \beta) \rightarrow Ve\Lambda\alpha\, X(\alpha, e|\alpha). \qquad (15)$$

But if one tries to reproduce the informal argument given in 10.4 for $\Lambda\alpha Vx$-continuity, we would have to use Abstr $(\alpha) = \alpha$ to justify this principle. See [M 1968], page 218, and this paper, 16.4.

The only (weak) argument in favour of (15) is perhaps that continuous relationships are the only ones made possible explicitly, as is seen from the description of GC-sequences.

10.8.

Formal systems for intuitionistic analysis.

In [T 1968] a system CS for intuitionistic analysis with choice sequences is described. CS is based on four sorted intuitionistic predicate logic with equality, with numerical variables, vars. for constructive functions, choice sequences and vars. for elements of K, and contains the usual arithmetical axioms, axioms for K, abstraction operators with rules of λ-conversion, and the axiom of choice in the form

$$\Lambda x Va\, F(x, a) \rightarrow Vb\Lambda x\, F(x, \lambda y.b\{x, y\})$$

and the following axioms for choice sequences

$$F\alpha \rightarrow Ve[V\beta(e|\beta = \alpha) \wedge \Lambda\gamma\, F(e|\gamma)]$$
$$\Lambda\alpha Va\, F(\alpha, a) \rightarrow Ve\Lambda n(en \neq 0 \rightarrow Vb\Lambda\alpha \in n\, F(\alpha, b))$$
$$\Lambda\alpha V\beta\, F(\alpha, \beta) \rightarrow Ve\Lambda\alpha\, F(\alpha, e|\alpha).$$

For a more detailed description see [T 1968]; the axioms mentioned there are slightly different but equivalent.

CS contains a subsystem IDK (essentially consisting of the part of CS without choice variables).

For future reference we describe IDK here separately:
IDK is based on two sorted intuitionistic predicate logic with equality, with variables for numbers and constructive functions, abstraction operator and rule of λ-conversion, the usual arithmetical axioms (induction, successor-axioms, defining axioms for some primitive recursive functions like $+$, $\cdot$, $*$, j_1, j_2, pairing function), the axiom of choice in $\Lambda x Va$-form, and the axioms for a predicate constant K:

$$\bigwedge a[A_K(K, a) \rightarrow Ka]$$

$$\bigwedge a[A_K(Q, a) \rightarrow Qa] \rightarrow \bigwedge a[Ka \rightarrow Qa].$$

(Alternatively, instead of introducing a predicate constant K, one may introduce variables for elements of K).

The main result stated in [T 1968] is essentially this:
there exists a translation T of closed formulae of CS into formulae of IDK (more accurately: some conservative extension of IDK) such that

$$\vdash_{CS} A \leftrightarrow T(A)$$

for every closed formula A of CS. We can prove much more however:

$$\vdash_{CS} A \text{ iff } \vdash_{IDK} T(A)$$

can be proved finitistically.

This result establishes consistency for a considerable part of the theory of choice sequences. It enables us to enterpret the continuity axioms as <u>defining</u> a special interpretation of certain quantifier combinations.

Take e.g. (15): this may be read as an <u>explanation</u> of the quantifier combination $\bigwedge\alpha\bigvee\beta$. Another way of reading (15) is, that it expresses a restriction on possible proofs of $\bigwedge\alpha\bigvee\beta\, X(\alpha,\beta)$.

The unsatisfactory aspect of this interpretation of the theory of choice sequences is, that although we know it to be successful for CS, we do not know if this "interpretation" or "explanation" can be maintained for arbitrary intuitionistically acceptable extensions of CS.

§ 11. Spreads and a theory of real numbers

11.1.

In this section we introduce the notion of a spread, well-known from intuitionistic literature. The general notion of a spread, although not absolutely indispensable, is often very convenient. Moreover, it is necessary to understand the notion in order to be able to read literature on "traditional intuitionism". The notion of a spread-law has already been defined in the previous section. A generalization of the definition 10.6.1 is obtained if we replace **a** by another kind of sequence.

11.1.1. Definition. A complementary (lawlike) mapping ξ of a tree or spread-law associates objects of a species S to the positive natural numbers (or, equivalently, to $\{n : an \neq 0 \wedge n \neq 0\}$ if a is the spread-law considered). The notion of a spread may be generalized by admitting non-lawlike mappings for a and/or for ξ.

A pair $\langle a, \xi\rangle$ such that $\underline{Spr}(a)$, ξ a complementary mapping into a species S is called a dressed spread. When we take for ξ a mapping into N such that $\xi\langle x_0, \ldots, x_u\rangle = x_u$, then $\langle a, \xi\rangle$ is a naked or undressed spread. When we talk about a spread a(with $\underline{Spr}(a)$) we mean the undressed spread associated with a.

$\langle a, \xi\rangle$ is said to be a subspread of $\langle a', \xi'\rangle$ if $X = \{n : an \neq 0\} \subseteq \{n : a'n \neq 0\}$ and if $\xi | (X - \{0\}) = \xi' | (X - \{0\})$.

A sequence $\chi \in (N)S$ is said to be an element of the spread $\langle a, \xi\rangle$ (notation $\chi \in \langle a, \xi\rangle$) if

$$V\chi' \in (N)N[\Lambda n(a\bar{\chi}'n \neq 0) \wedge \lambda x.\ \xi\bar{\chi}'(n+1) = \chi].$$

With a complementary mapping ξ a mapping ξ^* is associated defined by

$$\xi^*\chi = \langle \xi\bar{\chi}(n+1)\rangle_n = \lambda x.\xi\bar{\chi}(n+1) \qquad (\chi \in (N)N).$$

11.1.2. Definition. A species X is said to be represented by a spread with an equivalence relation $\sim$ (represented by $\langle a, \xi, \sim\rangle$) if $\langle a, \xi\rangle$ is a spread, and if S^*, the species of equivalence classes w.r.t. $\sim$ can be mapped bi-uniquely onto X. If this mapping is lawlike, the representation is called lawlike.

11.1.3. Definition. A spread with a finitary spread-law is called a fan or finitary spread.

11.2.

Let $\langle r_n\rangle_n$ be any given lawlike enumeration of the rational numbers, and let X be an extensional species of sequences of $(N)N$.

Let Rng X denote the collection of real number generators $\langle r_{\chi n}\rangle_n$, $\chi \in X$. The equivalence classes of Rng X with respect to the relation $\sim$ for real number generators (as defined in § 6) constitute a species of reals relative to X, $\langle r_n\rangle_n$: Re X.

If X is closed under composition with lawlike functions, i.e. if

$$\bigwedge \chi \bigwedge a(\chi \in X \to \lambda x.a\chi x \in X) \qquad (1)$$

then Rng X is independent of the particular lawlike enumeration of the rationals chosen.

For let $\langle r'_n\rangle_n$ be another such enumeration. Then there is a lawlike bi-unique mapping a of N onto N (supposing both enumerations to be without repetitions) such that $\bigwedge n(r'_n = r_{an})$. If $\langle r_{\chi n}\rangle_n$ is a real number generator, $\chi \in X$, then for $a^{-1} = b$, $\langle r_{\chi n}\rangle_n = \langle r'_{b\chi n}\rangle_n$, $b\chi \in X$. Conversely, if $\langle r'_{\chi n}\rangle_n$ is a r.n.g., $\chi \in X$, then $\langle r'_{\chi n}\rangle_n = \langle r_{a\chi n}\rangle_n$, $a\chi \in X$.

If we take for X the lawless sequences, condition (1) is not fulfilled. But even worse: $\langle r_{\chi n}\rangle_n$ can never be a real number generator for lawless χ, since

$$\bigwedge k \bigvee n \bigwedge m(|r_{\chi n} - r_{\chi(n+m)}| < 2^{-k})$$

is an a priori condition on all values of χ which makes χ non-lawless. (This is easily proved by the methods of section 9, and may be left as an exercise to the reader.)

On the other hand, (1) is fulfilled for lawlike and choice sequences.
For Re(X) not empty, real-valued (lawlike) functions from Re X into Re X are determined by lawlike operations $\Psi \in$ (Rng X) Rng X such that

$$\xi, \phi \in \text{Rng } X \wedge \xi \sim \phi \to \Psi\xi \sim \Psi\phi .$$

11.3.
There is a more general way of defining analysis of reals and real-valued functions, namely relative to a spread S and a species of sequences $X \subseteq (N)N$.

Let $S = \langle a, \phi\rangle$ be a spread, such that $\phi^*\chi$ is a real number generator for every $\chi \in a$. Rng(X, S) denotes the r.n.g's $\phi^*\chi$ for $\chi \in X$.

The species of equivalence classes w.r.t. equivalence between r.n.g's is denoted by Re(X, S). Real-valued functions relative to Re(X, S) are defined as in the case of Re(X).

We describe a spread S_0 which can be widely used. Let $\langle r_n \rangle_n$ be a fixed lawlike enumeration of rationals, as before, and let $\langle s(n, k, m) \rangle_m$ be an enumeration of all indexes of rational numbers s (without repetitions) such that $|r_n - s| < 2^{-k}$.

We define $S_0 = \langle a_0, \phi_0 \rangle$ by taking $a_0 = \lambda x.1$, and taking for ϕ_0:

$$\phi_0 \langle k_0 \rangle = r_{k_0}$$

$$\phi_0 \langle k_0, \ldots, k_{m-1} \rangle = r_n \rightarrow \phi_0 \langle k_0, \ldots, k_{m-1}, k \rangle = r_{s(n,m,k)} .$$

Pictorially represented, S_0 looks like

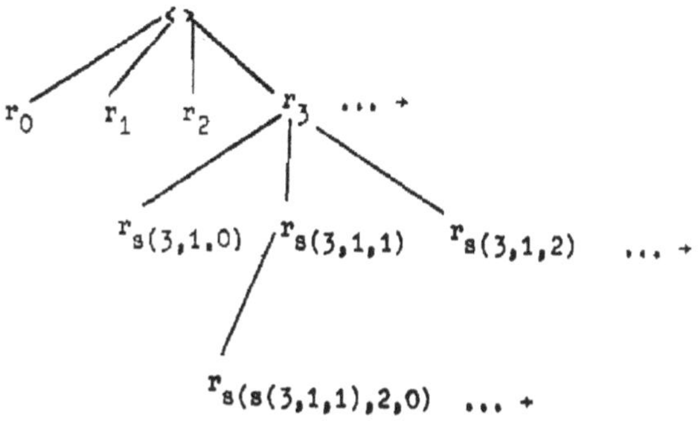

If $\langle s_n \rangle_n \in \langle a_0, \phi_0 \rangle$, then $\langle s_n \rangle_n$ is a real number generator, since

$$|s_{n+m} - s_n| \leq |s_{n+m} - s_{n+m-1}| + \ldots + |s_{n+1} - s_n| < 2^{-n-m+1} + \ldots + 2^{-n-1} + 2^{-n} < 2.2^{-n}. \quad (2)$$

S_0 does not contain all possible real number generators, since we can find an a priori bound on the rate of convergence, valid for every $\chi \in S_0$, as will be clear from (2).

It is possible to construct a spread S_1 which contains all real number generators. We leave the construction of S_1 as an excercise.

With respect to S_0 we can talk about $\mathbf{Rng}(LS, S_0)$ and $Re(LS, S_0)$. But once again lawless sequences do not produce a satisfactory theory of reals and real-valued functions; for example, $0 \notin Re(LS, S_0)$! And if we look for real-valued functions, matters become (if possible) worse. Because of the lawlessness, if $\langle r_n\rangle_n$, $\langle s_n\rangle_n \in \mathrm{Rng}(LS, S_0)$, then

$$\langle r_n\rangle_n \sim \langle s_n\rangle_n \leftrightarrow \bigwedge n(r_n = s_n)$$

since if $\langle r_n\rangle_n \neq \langle s_n\rangle_n$, $\langle r_n\rangle_n \sim \langle s_n\rangle$ would impose a condition of relative dependence on the underlying lawless sequences of $(N)N$, contradicting their lawlessness.

Hence a lawlike real valued function must be representable by a predicate $X(\alpha,\beta)$ $(\alpha,\beta \in LS)$ such that $\bigwedge\alpha\bigvee!\beta\ X(\alpha,\beta)$ which implies $\bigwedge\alpha X(\alpha,\alpha)$, so the corresponding real valued function must be the identity: $fx = x$ for all $x \in \mathrm{Re}(LS, S_0)$.

If we admit extra parameters in X (so the real valued function f defined by X need not be lawlike), the situation does not improve: in 9.9 we derived (α,β,γ LS-variables)

$$\bigwedge\alpha\bigvee!\beta X(\alpha,\beta,\gamma) \rightarrow \bigwedge\alpha(X(\alpha,\alpha,\gamma) \vee X(\alpha,\gamma,\gamma))$$

hence if f_{x_0} is the function defined by X, and x_0 the real number corresponding to γ in $\mathrm{Re}(LS, S_0)$, then

$$\bigwedge x(f_{x_0}x = x \vee f_{x_0}x = x_0) .$$

11.4.
If we restrict the spread S_0 described in 11.3 to a spread S_2 such that $\langle s_n\rangle_n \in S_2 \leftrightarrow \bigwedge n(|s_n| < 1)$, then S_2 represents $[-1,1]$ in the sense that any $\chi \in S_2$ is a r.n.g. for a number of $[-1,1]$.

For $[-1,1]$ we define yet another representation by a fan $S_3 = \langle a_3,\phi_3\rangle$.

a_3 is given by

$$\begin{cases} a_3 0 \neq 0 \\ a_3 \langle x_0,\ldots, x_v\rangle \neq 0 \leftrightarrow \bigwedge i \leq v(x_i \leq 2). \end{cases}$$

ϕ_3 is given by

$$\begin{cases} \phi_3 \langle k_0\rangle = (k_0-1)2^{-1} \\ \phi_3 \langle k_0,\ldots, k_n\rangle = \phi_3 \langle k_0,\ldots, k_{n-1}\rangle + (k_n-1)2^{-n-1} \end{cases} .$$

In a picture:

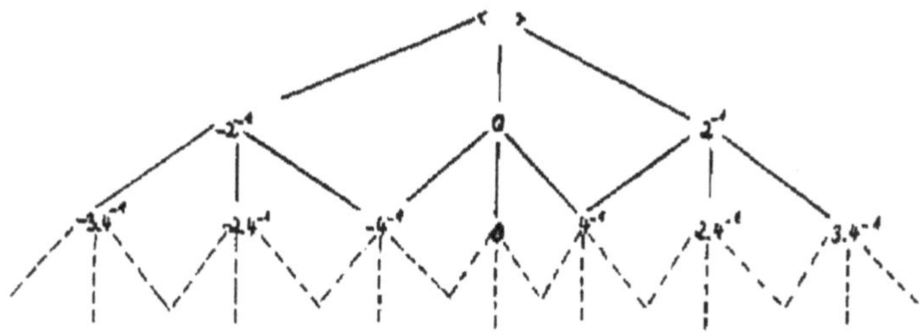

It is not hard to see that every $\chi \in S_3$ is a r.n.g. for an $x \in [-1,1]$.

We remark that $\mathrm{Re}(X)$, $\mathrm{Re}(S_0,X)$ are equivalent for X = species of lawlike sequences or X = species of choice sequences in the sense that there are lawlike operations Ψ, Ψ' such that

$$\Psi \in (\mathrm{Rng}\ X)\ \mathrm{Rng}(S_0,\ X),\ \Psi' \in (\mathrm{Rng}(S_0,\ X))\ \mathrm{Rng}\ X, \text{ and}$$

$$\Lambda \chi \in \mathrm{Rng}\ X(\Psi\chi \sim \chi),\ \Lambda\chi \in \mathrm{Rng}(S_0,\ X)(\Psi'\chi \sim \chi).$$

For Ψ' we may take the identity.

The existence of Ψ is proved by showing

$$\Lambda\chi \in \mathrm{Rng}(X)\ V\chi' \in \mathrm{Rng}(S_0,\ X)(\chi \sim \chi').$$

(By a selection principle this implies the existence of a lawlike operator, not necessarily continuous.) We take the case of lawlike sequences.

$\chi \equiv \langle s_n \rangle_n \in \mathrm{Rng}\ X$ implies

$$\Lambda k V i \Lambda n \Lambda m(|s_{i+n} - s_{i+m}| < 2^{-k}).$$

Hence $V b \Lambda k \Lambda n \Lambda m(|s_{bk+n} - s_{bk+m}| < 2^{-k})$.

Define c by

$$\begin{cases} c0 = b0 \\ c(k+1) = c(k) + b(k+1). \end{cases}$$

It follows that $\langle s_{ck} \rangle_k \in \mathrm{Rng}(S,\ X)$, since $|s_{c(k+1)} - s_{ck}| =$ $|s_{b(k)+i} - s_{b(k)+j}| < 2^{-k}$ for certain i, j, hence $\Lambda k |s_{c(k+1)} - s_{ck}| < 2^{-k}$.

Likewise one easily proves an equivalence for $Re(S_2, X)$, $Re(S_3, X)$, for $X = GC$ or X = species of lawlike sequences.

Remark. It follows from our experiences with lawless sequences that it is not entirely trivial to prove such equivalences!

11.5.
The homogeneous character of S_3 is expressed by:

Theorem. The subspread $S = \langle a, \phi \rangle$ of S_3, defined by:
$am \neq 0 \leftrightarrow (m \leq^{0} \langle k_0, \ldots, k_n \rangle \vee \langle k_0, \ldots, k_n \rangle <^{0} m) \wedge a_3 m \neq 0$,
ϕ obtained by restricting ϕ_3 to $\{m : am \neq 0 \wedge m \neq 0\}$ represents
$[\phi \langle k_0, \ldots, k_n \rangle - 2^{-n-1}, \ \phi \langle k_0, \ldots, k_n \rangle + 2^{-n-1}] \cap$ choice reals.

Proof. Let us extend ϕ^* to finite sequences by stipulating

$$\phi^* \bar{b}(n + 1) = (\overline{\phi^* b})(n + 1) \quad \text{for all } b, n.$$

Then S contains exactly the $\chi \in S_3$ with an initial segment $\phi^* \langle k_0, \ldots, k_n \rangle$.

We give the proof of our theorem for the case $S = S_3$, in such a way that it can be straightforwardly adapted to obtain a proof for any S determined by $\langle k_0, .., k_n \rangle$.

We give the proof for lawlike reals, for simplicity in notation; but in the case of choice reals the proof is completely analogous. Let x be any lawlike real number in $[-1,1]$. Then

$$\bigwedge k \bigvee n (|x - n.2^{-k}| < 5.2^{-k-3} \wedge |n| \leq 2^k).$$

Hence

$$\bigvee b \bigwedge k (|x - b(k).2^{-k}| < 5.2^{-k-3} \wedge |b(k)| \leq 2^k), \text{ so}$$

$$|2^{-k}.b(k) - 2^{-k-1}. b(k+1)| \nleq |x - b(k). 2^{-k}| + |x - b(k+1).2^{-k-1}|$$
$$\nleq 5.2^{-k-3} + 5.2^{-k-4} = 15.2^{-k-4} < 2^{-k}.$$

Therefore $|2b(k) - b(k+1)| < 2$, so

$$|2b(k) - b(k+1)| \leq 1.$$

Now we define c by

$$\begin{cases} |b(k)| < 2^k \rightarrow c(k) = b(k) \\ b(k) = 2^{-k} \rightarrow c(k) = b(k)-1 \\ b(k) = -2^k \rightarrow c(k) = b(k)+1. \end{cases}$$

If $|b(k)| < 2^k$, $|b(k+1)| < 2^{k+1}$, we have $|2\ c(k) - c(k+1)| \leq 1$.
If $b(k+1) = 2^{k+1}$, then $b(k) = 2^k$; hence again $|2\ c(k) - c(k+1)| \leq 1$.
If $b(k) = 2^k$, then $b(k+1) = 2^{k+1}$ or $b(k+1) = 2^{k+1} - 1$. In both cases, $|2c(k) - c(k+1)| \leq 1$. Likewise for $b(k) = -2^k$ or $b(k+1) = -2^{k+1}$. Therefore

$$|2\ c(k) - c(k+1)| \leq 1.$$

Now x is represented by $\langle c(k+1).\ 2^{-k-1}\rangle_k \in S_3$.

11.6.

Let α,β,γ denote choice sequences. Let $S_\alpha = \langle\lambda x.1,\xi\rangle$, where ξ is defined by $\xi 0 = 0$, $\bigwedge x(\xi(x+1) = \alpha x)$, α an arbitrary but fixed choice sequence.

We define $S'' = \langle\lambda x.1,\phi\rangle$ by stipulating

$$\phi(n * \langle x\rangle) = j_2 x\ .$$

S_α is a subspread of S'' in the sense described below.
We define

$$\begin{cases} \beta 0 = 1,\ \beta\langle x_0,\ldots,x_u\rangle = 1 \leftrightarrow \\ \alpha\langle j_1x_0\rangle = j_2x_0 \wedge \alpha\langle j_1x_0,\ j_1x_1\rangle = j_2x_2 \wedge \ldots \wedge \\ \wedge\ \alpha\langle j_1x_0,\ldots,j_1x_u\rangle = j_2x_u,\ \beta n = 0 \text{ otherwise.} \end{cases}$$

β is obtained from α by a continuous operation; hence β is a choice sequence.

β has all the properties of a spread-law, for $\beta(n * m) \neq 0 \rightarrow \beta n \neq 0$ follows immediately from the definition, and if $\beta\langle x_0,\ldots,x_u\rangle \neq 0$, and we take any x, then $\beta\langle x_0,\ldots,x_u,y\rangle \neq 0$ for $y = \{x,\ \alpha\langle j_1x_0,\ldots,j_1x_u,\ x\rangle\}$. Now if $X = \{n : \beta n \neq 0\}$, then $\langle\beta,\phi|X\rangle$ contains exactly the same choice sequences as S_α.

The more general case, where $S_\alpha = \langle a,\ \xi\rangle$, a an arbitrary lawlike spread law, ξ defined as before, is treated likewise, with as a result:

<u>Theorem</u>. To every $S_\alpha = \langle a,\ \xi\rangle$, with $\xi 0 = 0$, $\bigwedge x(\xi(x+1) = \alpha x)$ α a choice sequence, a spread S with a choice-spread law and a lawlike complementary law may be found such that $\{\beta : \beta \in S_\alpha\} = \{\beta : \beta \in S\}$.

<u>Remark</u>. This theorem expresses the essential content of Brouwer's often misunderstood paper [Br 1942 A].

Conversely, if one has a spread $\langle\alpha,a\rangle$, α a spread law, $a \in (N)N$ a lawlike complementary law, then the species of choice sequences contained in $\langle\alpha,a\rangle$ may be represented as $\langle\lambda x.1,\beta\rangle$ such that for all γ $a^*(\Gamma_\alpha\gamma) = \beta^*(\gamma)$, where Γ_α is defined like Γ_a in 10.6.3.

§ 12. Topology; separable metric spaces

12.1.

In order to be able to present some mathematical applications of the principles for choice sequences, we discuss in this section the introduction of separable metric spaces in intuitionistic mathematics. α,β,γ are variables for choice sequences.

12.1.1. Definition. A metric space is a pair $\langle X,\rho\rangle$, X a species, ρ a metric on X, i.e. a (lawlike) real valued function on $X \times X$ such that for all $x,y,z \in X$

a) $\rho(x,y) \nless 0$, $\rho(x,y) = 0 \leftrightarrow x = y$,

b) $\rho(x,y) = \rho(y,x)$

c) $\rho(x,y) \nless \rho(x,z) + \rho(z,x)$.

The values of ρ are supposed to be choice reals.

A metric automatically induces an apartness relation given by $\rho(x,y) > 0 \leftrightarrow x \# y$.
$U(\varepsilon,p) = \{q : \rho(p,q) < \varepsilon\}$.

12.1.2. Definition. A sequence of points $\langle p_n\rangle_n$ is said to be dense in a space $\langle X,\rho\rangle$ if

$$\wedge p \in X \wedge k \vee n(\rho(p,p_n) < 2^{-k}).$$

A sequence of points $\langle p_n\rangle_n$ is called fundamental if

$$\wedge k \vee n \wedge m(\rho(p_n,p_{n+m}) < 2^{-k}.$$

Two fundamental sequences are said to be equivalent (notation $\langle p_n\rangle_n \sim \langle q_n\rangle_n$) if

$$\wedge k \vee n \wedge m(\rho(p_{n+m},q_{n+m}) < 2^{-k}.$$

p is said to be the limit of a sequence $\langle q_n\rangle_n$

(notation $\lim_n q_n = p$ or $\lim_{n\to\infty} q_n = p$) if

$$\wedge k \vee m \wedge n(\rho(p,q_{n+m}) < 2^{-k}).$$

12.1.3. Definition. A space $\langle X,\rho\rangle$ is said to be a separable metric space (with respect to choice sequences) if there is a lawlike sequence $\langle p_n\rangle_n \subseteq X$ ($\langle p_n\rangle_n$ is called a basis for $\langle X,\rho\rangle$) such that

$$\bigwedge p\in X \bigvee \alpha(\lim_{n\to\infty} p_{\alpha n} = p).$$

12.1.4. Remark. We might have defined "separable" with respect to other notions of sequence also; but in the sequel we shall consistently assume "separable" to be defined relative to choice sequences.

12.1.5. Definition. A separable metric is said to be complete (with respect to choice sequences) if for the lawlike $\langle p_n\rangle_n$ of the previous definition, every fundamental $\langle p_{\alpha n}\rangle_n$ converges to a point $p\in X$.

12.1.6. Remark. Let $\langle p_n\rangle_n$ and $\langle q_n\rangle_n$ be lawlike sequences, both dense in X. Then $\bigwedge k \bigwedge n \bigvee m(\rho(p_m,q_n) < 2^{-k})$, hence $\bigvee a \bigwedge k \bigwedge n(\rho(p_{a\{n,k\}}, q_n) < 2^{-k})$. Let $\langle q_{\alpha n}\rangle_n$ be a fundamental sequence; then $\langle q_{\alpha n}\rangle_n$ is equivalent with $\langle p_{a\{\alpha n,n\}}\rangle_n$, since $\rho(q_{\alpha n},p_{a\{\alpha n,n\}}) < 2^{-n}$. Since $\lambda n.a\{\alpha n,n\}$ is a choice sequence, say β, we have $\langle q_{\alpha n}\rangle_n \sim \langle p_{\beta n}\rangle_n$. The correspondence is lawlike.

In case the sequence $\langle q_n\rangle_n$ is lawlike relative to a choice parameter, the correspondence is lawlike relative to the same choice parameter.

12.2.

A basis $\langle p_n\rangle_n$ for a separable metric space $\langle X,\rho\rangle$ is not necessarily discrete ($\langle p_n\rangle_n$ is called discrete if $\bigwedge n \bigwedge m(p_n \# p_m \vee p_n = p_m)$). We can prove the following theorem however.

Theorem. Let $\langle X,\rho\rangle$ be a separable metric space with a basis $\langle p_n\rangle_n$. Then we can find a sub-sequence $\langle p_{an}\rangle_n$ which is a basis for $\langle X,\rho\rangle$ and which is discrete.

Proof. Let b be a lawlike function such that

$$|b(n,m,k).2^{-k} - \rho(p_n,p_m)| < 2^{-k}.$$

Now we define a by recursion, and we put

$$a0 = 0.$$

We want to achieve

$$\begin{cases}\rho(p_0,p_1) < 2^{-2} \to a1 = 0,\\ \rho(p_0,p_1) \nless 2^{-1} \to a1 = 1.\end{cases}$$

This can be done by stipulating

$$b(0, 1, 2) < 2 \rightarrow a1 = 0$$
$$a1 = 1 \text{ otherwise.}$$

For $|b(0, 1, 2)2^{-2} - \rho(p_0, p_1)| < 2^{-2}$, hence $\rho(p_0, p_1) < 2^{-2}$ implies $b(0, 1, 2) < 1 + 4\rho(p_0, p_1) < 2$. Conversely, $\rho(p_0, p_1) \nless 2^{-1}$ implies $(b(0, 1, 2) > | 4\rho(p_0, p_1) - 1| \nless 1$, i.e. $b(0, 1, 2) > 1$.

Now suppose $a0, \ldots, ac(k-1)$ to be constructed after the kth step, such that $\bigwedge i(i \leq c(k-1) \rightarrow ai \leq k-1)$.

Let $\langle j_0, \ldots, j_n \rangle$ be an ordering of the elements of $\{a0, \ldots, ac(k-1)\}$ such that $n_1 < n_2 \rightarrow j_{n_1} < j_{n_2}$. Let $\langle j_{n+1}, \ldots, j_k \rangle$ be an ordering of the elements of $\{0, \ldots, k\} - \{j_1, \ldots, j_n\}$ such that $n_1 < n_2 \rightarrow j_{n+n_1} < j_{n+n_2}$ so $(j_k = k)$.

Construct $\langle j_1', \ldots, j_k' \rangle$ from $\langle j_0, \ldots, j_k \rangle$ as follows. Take $j_i' = j_i$ for $i \leq n$. If $j_0', \ldots, j_{m-1}'$ have been chosen, we want to choose j_m' such that

(a_k) If there exists an $i \leq m - 1$ such that $\rho(p_{j_i'}, p_{j_m}) < 2^{-k-1}$, then $j_m' = j_{m-1}'$.

(b_k) If for every $i \leq m-1$ $\rho(p_{j_i'}, p_{j_m}) \nless 2^{-k}$, then $j_m' = j_m$.

Denote $\inf \{b(j_m, j_i', k+1) : 0 \leq i \leq m\}$ by d.

Then we put

$$\begin{cases} d \leq 1 \rightarrow j_m' = j_{m-1}' \\ d > 1 \rightarrow j_m' = j_m' \end{cases}.$$

Suppose $\rho(p_{j_i'}, p_{j_m}) < 2^{-k-1}$ for an $i \leq m-1$. Then $|b(j_m, j_i', k+1)\, 2^{-k-1} - \rho(p_{j_i'}, p_{j_m})| < 2^{-k-1}$, therefore $b(j_m, j_i', k+1) < 1 + 1$, so $b(j_m, j_i', k+1) \leq 1$, hence $d \leq 1$.

Conversely, suppose $\rho(p_{j_i'}, p_{j_m}) \nless 2^{-k}$ for every $i \leq m-1$. Then $b(j_m, j_i', k+1) > 1$ for all $i \leq m-1$, hence $d > 1$. This shows that our construction of j_m' satisfies (a_k), (b_k).

Finally we take

$$a(c(k-1) + i) = j'_{n+i} \quad \text{for} \quad 1 \leq i \leq k-n.$$

Apparently $c(k) = c(k-1) + (k-n)$.

Now we prove the discreteness of $\langle p_{an}\rangle_n$ by proving the discreteness of $\langle p_{a0}, \ldots, p_{ac(k)}\rangle$ for every k. Suppose the discreteness of $\langle p_{a0}, \ldots, p_{ac(k-1)}\rangle$ to be proved. From the fact that (a_k), (b_k) are satisfied at every step, it follows that a $j_m (j_m \notin \{j'_0, \ldots, j'_{m-1}\})$ is included in $\{j'_0, \ldots, j'_k\}$ only in case p_{j_m} lies apart from every element of $\{p_{j'_0}, \ldots, p_{j'_{m-1}}\}$. Therefore $\langle p_{a0}, \ldots, p_{ac(k)}\rangle$ is discrete.
Thus we obtain the desired conslusion by ordinary induction.

There remains to be shown that $\langle p_{an}\rangle_n$ is a basis. In view of 12.1.6, it is sufficient to show $\langle p_{an}\rangle_n$ to be dense in the space. To see this, we remark that for every $i < k$ either $p_i \in \{p_{a0}, \ldots, p_{ac(k)}\}$, or there exists an $\ell (0 \leq \ell \leq c(k))$ such that $\rho(p_i, p_{a\ell}) < 2^{-k+1}$ (for if $i \notin \{j'_1, \ldots, j'_k\}$, $i \leq k$ at the kth step, there exists a j'_m such that $\rho(p_{j'_m}, p_i) < 2^{-k+1}$, $m \leq k$).

So if we take $\rho(p, p_n) < 2^{-k-1}$, then we can effectively find a $p_{a\ell}$, $\ell \leq c(k+2)$, such that $\rho(p_n, p_{a\ell}) < 2^{-k-1}$, hence $\rho(p, p_{a\ell}) < 2^{-k}$.

Q.e.d. ([T 1966], 2.2.1)

12.3.

12.3.1. <u>Theorem</u>. Every complete separable metric space $\Gamma = \langle X, \rho\rangle$ can be represented by a spread.

<u>Proof</u>. Let $\langle p_n\rangle_n$ be a discrete basis for Γ, and let $\langle s(k, n, m)\rangle_m$ be a lawlike enumeration of all indexes i (possibly with repetitions) such that $\rho(p_n, p_i) < 2^{-k}$. We may suppose s to be chosen such that

$$p_n = p_m \rightarrow \Lambda k \Lambda l\, (s(k, n, l) = s(k, m, l)).$$

(This is not essential but technically convenient.) Now we define $S = \langle a, \phi\rangle$ by taking $a = \lambda x.1$, and fixing ϕ by

$$\begin{cases} \phi\langle k_0\rangle = p_{k_0} \\ \phi\langle k_0, \ldots, k_{n-1}\rangle = p_m \rightarrow \phi\langle k_0, \ldots, k_{n-1}, k\rangle = p_{s(n, m, k)} \end{cases}.$$

Then just as in 11.3, S represents Γ.

12.3.2. Lemma. Let $\Gamma = \langle X, \rho \rangle$ be a complete metric space, then the spread representation indicated in the proof of 12.3.1 has the following property:

$$\bigwedge x \in X \bigvee \alpha (\lim \phi^{*} \alpha = x \wedge \bigwedge k (U(2^{-k}, x) \subseteq W_{\bar{\alpha}k}))$$

where

$$W_n = \{y : \bigvee \beta \in n (\lim \phi^{*} \beta = y)\}.$$

Proof. Take a certain $x \in X$ and let y be such that $\rho(x, y) < 2^{-k-2}$. Construct $\langle q_n \rangle_n \subseteq \langle p_n \rangle_n$ such that $\bigwedge n (\rho(q_n, x) < 2^{-n-2})$. Then apparently

$$\bigwedge n (\rho(q_n, q_{n+1}) \nleq \rho(q_n, x) + \rho(q_{n+1}, x) < 2^{-n})$$

so $\langle q_n \rangle_n \in S$. Let $\phi \langle v_0, \ldots, v_{k-1} \rangle = q_{k-1}$. Clearly,

$$\rho(y, q_{k-1}) \nleq \rho(x, y) + \rho(x, q_{k-1}) < 2^{-k-2} + 2^{-k-1} < 3.2^{-k-2} .$$

Then we can find $\langle q'_n \rangle_n \subseteq \langle p_n \rangle_n$ such that $\lim \langle q'_n \rangle_n = y$, $\bigwedge n (\rho(q'_n, y) < 2^{-n-2})$, hence $\rho(q'_k, q_{k-1}) < 2^{-k}$.
Then the sequence $\langle q_0, q_1, \ldots, q_{k-1}, q'_k, q'_{k+1}, q'_{k+2}, \ldots \rangle \in S$, as is easily verified. This sequence is equal to $\phi^{*} \beta$ for a suitable β, with $\bar{\beta} k = \langle v_0, \ldots, v_{k-1} \rangle$.
Therefore $y \in W_{\langle v_0, \ldots, v_{k-1} \rangle}$.

Q.e.d.

12.4.

12.4.1. Definition. A topological space is a pair $\langle V, \mathcal{T} \rangle$, $\mathcal{T}$ a species of subspecies of V such that

(a) $\emptyset \in \mathcal{T}$, $V \in \mathcal{T}$

(b) Finite intersections and arbitrary unions of elements of $\mathcal{T}$ again belong to $\mathcal{T}$.

This definition is exactly the same as one of the well known classical definitions.

12.4.2. Definition. f is called a continuous mapping from a topological space $\langle V, \mathcal{T} \rangle$ into a space $\langle V', \mathcal{T}' \rangle$ if f is a mapping $(V)V'$ such that

$$\bigwedge X \in \mathcal{T}' (f^{-1}[X] \in \mathcal{T}).$$

The definitions of homeomorphism, basis of open species, and neighbourhood are given accordingly.

A point p is said to be a closure point of $X \subseteq V$, if $\bigwedge W \in \mathcal{T} (p \in W \rightarrow \bigvee q (q \in X \cap W))$ (every neighbourhood of p contains a point of X).

X^-, the <u>closure</u> of X consists of the species of closure points of X.

So far, all things look very much the same as in classical topology. But we must be aware of the fact that classically equivalent definitions are not necessarily intuitionistically equivalent.

For example, the notion of weak continuity, characterized by:

$$\bigwedge X \subseteq V' \, (f^{-1}[X^-]^- = f^{-1}[X^-])$$

(the complete original or counterimage of a closed set is always closed) is classically equivalent with, but intuitionistically weaker than continuity ([T 1966], 2.1.8).

With a metric space $\langle V, \rho \rangle$ we may associate in the usual way a topological space $\langle V, \mathcal{T}(\rho) \rangle$ by putting

$$X \in \mathcal{T}(\rho) \leftrightarrow \bigwedge p \in X \bigvee \varepsilon > 0 \; \bigwedge q \in V \, (\rho(p, q) < \varepsilon \rightarrow q \in X).$$

Or if we put

$$U(\varepsilon, p) = \{q : \rho(p, q) < \varepsilon\}$$

then

$$X \in \mathcal{T}(\rho) \leftrightarrow \bigwedge p \in X \bigvee \varepsilon > 0 \; (U(\varepsilon, p) \subseteq X).$$

The ε-neighbourhoods $U(\varepsilon, p)$ constitute a basis for $\mathcal{T}(\rho)$.

Usually we shall not distinguish between $\langle V, \rho \rangle$ and $\langle V, \mathcal{T}(\rho) \rangle$ when the meaning is clear from the context.

12.5.

The results mentioned in this subsection may serve as an illustration of the more typical questions of intuitionistic topology.

12.5.1. <u>Definition</u>. A space $\Gamma = \langle V, \mathcal{T} \rangle$ is said to be <u>located compact</u> if Γ is complete, metrizable (i.e. $\mathcal{T} = \mathcal{T}(\rho)$, $\langle V, \rho \rangle$ complete, for a suitable ρ) and representable by a finitary spread.

12.5.2. <u>Definition</u>. A space $\Gamma = \langle V, \mathcal{T} \rangle$ is said to be <u>locally compact</u>, if Γ is complete, metrizable, and if every $p \in V$ possesses a located compact neighbourhood U (i.e. U is located compact in the relative topology $\mathcal{T}' = \{X \cap U : X \in \mathcal{T}\}$ induced by $\mathcal{T}$).

Closed intervals $[a, b] \subseteq \mathbb{R}$, $a \leq b$ are examples of located compact spaces;

$\mathbb{R}$, the species of (choice-) reals is an example of a locally compact space.

Arbitrary pointspecies in a topological space may be defined very nonconstructively; so one feels the need for a subclass of pointspecies for which some extra information is available. The notion of a located pointspecies turns out to be useful:

12.5.3. Definition. Let $\Gamma = \langle V, \mathcal{T}\rangle$ be a topological space. $X \subseteq V$ is said to be located, if

$$\bigwedge p \in V \bigwedge U \in \mathcal{T}(p \in U \rightarrow (\bigvee q(q \in U \cap X) \vee \bigvee W \in \mathcal{T}(p \in W \wedge W \cap X = \emptyset))).$$

12.5.4. Definition. A species $X \subseteq V$, $\langle V, \rho\rangle$ a metric space, is said to be metrically located if $\rho(x, X) = \inf\{\rho(x, y) : y \in X\}$ is defined for every $x \in V$.

The significance of the notion of locatedness becomes clear from the following result:

12.5.5. Theorem. Let $\Gamma = \langle V, \mathcal{T}(\rho)\rangle$ be a located compact space. Then an inhabited $X \subseteq V$ is located iff X is metrically located. (Proof e.g. in [T 1968 A], 3.14(a)).

So metrical locatedness is a topological notion. In general, this is not the case; e.g. $\mathbb{R}$ can be metrized in such a way that a certain located pointspecies cannot be proved to be metrically located. But the following theorem holds:

12.5.6. Theorem. To every locally compact space $\langle V, \mathcal{T}\rangle$ we can find a metric ρ such that $\mathcal{T} = \mathcal{T}(\rho)$, $\langle V, \rho\rangle$ complete, and every located inhabited species $X \subseteq V$ is metrically located with respect to ρ (the converse is trivial). A proof may be found in [T 1968 A], 4.6.

12.5.7. Definitions. Let $\langle V, \mathcal{T}\rangle$ be a topological space, $X \subseteq V$, $Y \subseteq V$.

Interior $X = \mathrm{Int}\, X = \{p : \bigvee \varepsilon\, U(\varepsilon, p) \subseteq X\}$.

$X \Subset Y \equiv_D \bigwedge p \in V(p \in Y \vee p \notin X)$

is the analogue of the classical relation $X^- \subseteq \mathrm{Int}\, Y$.

12.5.8. Theorem. Let $\langle V, \rho\rangle$ be a complete separable metric space. Then if X is located,

$$X \Subset Y \quad \text{iff} \quad X^- \subseteq \mathrm{Int}\, Y.$$

Proof. The proof is contained in [T 1968 B], and uses continuity principles for choice sequences in an essential way.

§ 13. Applications of the continuity principles and the fan theorem

13.1.

We start with some applications of the continuity principle for choice sequences in its simplest form

$$\Lambda\alpha V x\, X(\alpha, x, \alpha_0, \ldots) \rightarrow \Lambda\alpha V x V y \Lambda\beta(\bar{\alpha}y = \bar{\beta}y \rightarrow X(\beta, x, \alpha_0, \ldots)) .$$

Proof-theoretically, this principle is not an essential strengthening of analysis (as follows e.g. from [Ho,K 1967]) but its mathematical consequences are interesting and rather elegant.

13.1.1. Definition. Let $\Gamma = \langle V, \rho\rangle$, $\Gamma' = \langle V', \rho'\rangle$ and let f be a (lawlike) mapping from V into V'. f is said to be sequentially continuous if every converging sequence $\langle p_n\rangle_n \subseteq V$ with a limit p is mapped onto a converging sequence $\langle fp_n\rangle_n$ with a limit fp.

13.1.2. Theorem. Let Γ be a separable complete metric space, Γ' a separable metric space, f a mapping from Γ into Γ'. Then f is sequentially continuous.

Proof. See [T 1967]. This theorem is contained in the stronger theorem 13.1.6 given below.

13.1.3. Theorem. Let Γ' be a separable metric space, and let Γ'' be a metric space. If f is a (lawlike) mapping from Γ' into Γ'', and f is sequentially continuous, then f is continuous.

Proof. ([T 1967]). Let $\langle p_n\rangle_n$ be a lawlike sequence of points dense in $\Gamma' = \langle V, \rho\rangle$. Let $\Gamma'' = \langle W, \rho'\rangle$. Let x be an arbitrary point of Γ', and suppose $\lim \langle p_{\alpha n}\rangle_n = x$. Now we want to construct $\langle p_{\beta(n,i)}\rangle_i$ such that

$$\left.\begin{array}{l} \rho(p_{\beta(n,i)}, x) < 2.2^{-n} \text{ for every } i \\ \rho(p_i, x) \nmid 2^{-n} \rightarrow p_i \in \langle p_{\beta(n,j)}\rangle_j \end{array}\right\} \qquad (1)$$

The construction proceeds as follows.

From $\langle p_{\alpha n}\rangle_n$ we obtain effectively $\langle p_{\gamma n}\rangle_n$, $\lim \langle p_{\gamma n}\rangle_n = x$, such that

$$\Lambda n(\rho(p_{\gamma n}, x) < 2^{-n}).$$

Let $b(n, m, k)$ be a (lawlike) enumeration of rationals such that

$$\Lambda n \Lambda m \Lambda k\, |\rho(p_n, p_m) - b(n, m, k)| < 2^{-k}.$$

We have

$$b(i,\ \gamma(n+2),\ n+2) < 6.2^{-n-2} \vee b(i,\ \gamma(n+2),\ n+2) \geq 6.2^{-n-2}.$$

If $\rho(p_i,\ x) \ngtr 2^{-n}$, then $\rho(p_i,\ p_{\gamma(n+2)}) < 5.2^{-n-2}$; also $|\rho(p_i,\ p_{\gamma(n+2)}) - b(i,\ \gamma(n+2),\ n+2)| < 2^{-n-2}$, therefore $b(i,\ \gamma(n+2),\ n+2) < 6.2^{-n-2}$.

If $b(i,\ \gamma(n+2),\ n+2) < 6.2^{-n-2}$, then $\rho(p_i,\ p_{\gamma(n+2)}) < 7.2^{-n-2}$, so $\rho(p_i,\ x) < 8.2^{-n-2} = 2.2^{-n}$.

Therefore we may take for $\langle\beta\{n,i\}\rangle_i$ an enumeration of the j such that $b(j,\ \gamma(n+2),\ n+2) < 6.2^{-n-2}$; then (1) is satisfied. (β is obtained by a continuous operation from γ)).

Now we define a spread $\langle\lambda x.1,\phi\rangle$ by stipulating

$$\phi\langle i_0,\ldots,\ i_n\rangle = \langle p_{\beta\{0,i_0\}},\ldots,\ p_{\beta\{n,i_n\}}\rangle\ .$$

Clearly, for every α $'\phi^{*}\alpha$ is a fundamental sequence in Γ' converging towards x.

Furthermore, for any fixed natural number ν :

$$\Lambda\delta V n\Lambda m(\rho'(fp_{\beta\{n+m,\ \delta(n+m)\}},\ fx) < 2^{-\nu}),$$

hence with an application of the continuity principle we find, for any δ', n, n' such that

$$\Lambda\delta(\bar{\delta}'n' = \bar{\delta}n' \rightarrow \Lambda m(\rho'(fp_{\beta\{n+m,\ \delta(n+m)\}},\ fx) < 2^{-\nu})).$$

Without restriction we may suppose $n' = n$. From this it follows that

$$\Lambda i(\rho(p_i,\ x) < 2^{-n} \rightarrow \rho'(fp_i,\ fx) < 2^{-\nu}).$$

For whenever $\rho(p_i,\ x) < 2^{-n}$, there is a δ with $\bar{\delta}'n = \bar{\delta}n$, $\beta\{n,\ \delta n\} = i$.

Now take any y such that $\rho(x,y) < 2^{-n}$. Then there is a sequence $\langle p_{\delta n}\rangle_n \subseteq \langle p_i\rangle_i$, such that $\Lambda m(\rho(p_{\delta m},\ x) < 2^{-n})$, $\lim_n p_{\delta n} = y$.
f is sequentially continuous, hence for a suitable k $\rho'(fp_{\delta k},\ fy) < 2^{-\nu}$.
Therefore

$$\rho'(fy,\ fx) \ngtr \rho'(fp_{\delta k},\ fy) + \rho'(fp_{\delta k},\ fx) < 2^{-\nu} + 2^{-\nu} = 2.2^{-\nu}$$

and thus

$$\Lambda y(\rho(y,\ x) < 2^{-n} \rightarrow \rho'(fy,\ fx) < 2.2^{-\nu})\ .$$

q.e.d.

13.1.4. Theorem. Let $\Gamma = \langle V, \mathcal{T}(\rho)\rangle$ be a separable, complete, metrizable space. Let Γ be covered by a species $\mathfrak{A}$ of pointspecies; then Γ is also covered by the species of interiors of these pointspecies.

Proof. Let us use W as a variable for pointspecies.
Then we have $\Lambda x \in V\, \mathsf{V} W(W \in \mathfrak{A} \wedge x \in W)$.

By 12.3.1 there is a representation $\langle \lambda x.1, \phi\rangle$ for Γ.
Hence we also have:

$$\Lambda \alpha\, \mathsf{V} W(W \in \mathfrak{A} \wedge \lim \phi^* \alpha \in W).$$

Now we apply (13) of 10.7 and we obtain

$$\Lambda \alpha \mathsf{V} n \mathsf{V} W \Lambda \beta(\bar{\alpha}n = \bar{\beta}n \rightarrow \lim \phi^* \alpha \in W \wedge W \in \mathfrak{A}).$$

We then use lemma 12.3.2.
Take any x, and find α such that

$$\lim \phi^* \alpha = x \wedge \Lambda k(U(2^{-k}, x) \subseteq W_{\bar{\alpha}k}).$$

We have an n and a $W \in \mathfrak{A}$ such that

$$\Lambda \beta(\bar{\alpha}n = \bar{\beta}n \rightarrow \lim \phi^* \beta \in W).$$

Hence $U(2^{-n}, x) \subseteq W$, so $x \in \operatorname{Int} W$.
This proves $\Lambda x \in V\, \mathsf{V} W(W \in \mathfrak{A} \wedge x \in \operatorname{Int} W)$.

13.1.5. Remark. In case $\mathfrak{A}$ was a subspecies of N, it would have been sufficient to use (3) of § 10 instead of (13).

We remark that the lawlikeness of $\mathfrak{A}$ and the elements of $\mathfrak{A}$ is essential.

For consider the covering by species of one element; then the interiors do not cover Γ! The reason is that arbitrary points are not lawlike. The theorem may be generalized however to the case where the elements of $\mathfrak{A}$ contain choice parameters from a fixed finite collection.

From 13.1.4 one also obtains

13.1.6. Theorem. Let Γ be a complete, separable, metric space, and let Γ'' be a separable metric space; then any mapping from Γ into Γ'' is continuous.

Proof. Either we may combine 13.1.2 and 13.1.3, or we deduce the theorem directly form 13.1.4, by looking for any fixed natural number ν at the coverings $\{W_{\nu,i} : i \in N\}$, defined as follows. Let $\langle p_n \rangle_n$ be a basis for $\Gamma'' = \langle Y, \rho\rangle$; then

$$W_{\nu,1} = \{y : y \in \Gamma \wedge \rho(p_1, fy) < 2^{-\nu}\} .$$

Then an application of 13.1.4 readily yields continuity of f.

13.2.

Intermediate in strength between (13) of § 10 and the corresponding stronger form

$$\bigwedge\alpha \bigvee X^\sigma Y(\alpha, X^\sigma) \rightarrow \bigvee e \bigwedge n(en \neq 0 \rightarrow \bigvee X^\sigma \bigwedge \alpha \in n Y(\alpha, X^\sigma))$$

we have

$$\bigwedge\alpha \bigvee X^\sigma Y(\alpha, X^\sigma) \rightarrow \bigvee a[\bigwedge n(an \neq 0 \rightarrow \bigvee X^\sigma \bigwedge \alpha \in n Y(\alpha, X^\sigma)) \wedge \bigwedge\alpha\bigvee n(a\bar{\alpha}n \neq 0)] .$$

If we apply this principle instead of the weak form of continuity, one obtains an intuitionistic version of Lindelöfs theorem:

13.2.1 <u>Theorem</u>. Let $\Gamma = \langle V, \mathcal{H}(\rho)\rangle$ be a separable, complete, metric space, and let Γ be covered by a species $\mathfrak{A}$ of pointspecies. Then there exists an enumerable subspecies $\mathcal{B} \subseteq \mathfrak{A}$ such that $\{\mathrm{Int}\ W : W \in \mathcal{B}\}$ is a covering for Γ.

<u>Proof</u>. The proof can be given by remarking that if

$$\bigwedge\alpha \bigvee W \in \mathfrak{A} \lim \phi^* \alpha \in W$$

and if a is such that $\bigwedge\alpha\bigvee n(a\bar{\alpha}n \neq 0)$, $\bigwedge n(an \neq 0 \rightarrow \bigvee W \in \mathfrak{A} \bigwedge \alpha \in n(\lim \phi^* \alpha \in W))$, then there is a species $W^*[n]$ with parameter n such that

$$\bigwedge n(an \neq 0 \rightarrow W^*[n] \in \mathfrak{A} \wedge \bigwedge \alpha \in n(\lim \phi^* \alpha \in W^*[n])).$$

The collection $\mathcal{B} = \{W^*[n] : an \neq 0\}$ covers Γ; then apply 13.1.4.

13.3.

We finish this section with applications of the fan theorem. Very well known is

13.3.1. <u>Theorem</u> ([H 1966], 3.4.3). A real-valued (lawlike) function on $[-1, 1]$ is uniformly continuous.

<u>Proof</u>. Let us use the spread S_3 described in 11.4, with the theorem proved in 11.5. $S_3 = \langle a_3, \phi_3 \rangle$, f is a real-valued lawlike function on $[-1, 1]$.

Let ν be a fixed natural number, and let S be the undressed spread a_3. For any fixed $\nu \in N$:

$$\bigwedge \alpha \in a_3 \bigvee m |f \lim \phi^* \alpha - m.2^{-\nu}| < 2^{-\nu}$$

or

$$\bigwedge\alpha\bigvee m |f \lim \phi^* \Gamma_{a_3} \alpha - m.2^{-\nu}| < 2^{-\nu}.$$

Applying the fan theorem yields:

$$\bigvee n \bigwedge \alpha \bigvee m \bigwedge \beta(\overline{\Gamma_{a_3}\alpha}\, n = \overline{\Gamma_{a_3}\beta}\, n \rightarrow |f \lim \phi^* \Gamma_{a_3} \beta - m.2^{-\nu}| < 2^{-\nu})$$

or equivalently

$$\bigvee n \bigwedge \alpha \in a_3 \bigvee m \bigwedge \beta \in a_3 (\bar{\alpha}n = \bar{\beta}n \to |f \lim \phi^* \beta - m.2^{-\nu}| < 2^{-\nu}).$$

Suppose n to be such that $\bigwedge \alpha \in a_3 \bigvee m \bigwedge \beta \in a_3 (\bar{\alpha}n = \bar{\beta}n \to |f \lim \phi^* \beta - m.2^{-\nu}| < 2^{-\nu})$, and let x, y be such that $|x-y| < 2^{-n-1}$. Then we can find $\alpha, \beta \in a_3$ such that $\lim \phi^* \alpha = x$, $\lim \phi^* \beta = y$, $\bar{\alpha}n = \bar{\beta}n$, since it follows from 11.5 that if we define

$$W_{\bar{\alpha}n} = \{\lim \phi^* \beta : \bar{\beta}n = \bar{\alpha}n \wedge \beta \in a_3\}$$

then $\{W_{\bar{\alpha}n} : \alpha \in a_3\}$ is exactly the collection

$$\{[(k-1)\, 2^{-n-1}, (k+1) 2^{-n-1}] : -2^{n+1} < k < 2^{n+1}\} .$$

Let m be such that

$$\bigwedge \gamma \in a_3 (\bar{\alpha}n = \bar{\gamma}n \to |f \lim \phi^* \gamma - m.2^{-\nu}| < 2^{-\nu})$$

then

$$|fx - fy| = |f \lim \phi^* \alpha - f \lim \phi^* \beta| \leq$$
$$|f \lim \phi^* \alpha - m.2^{-\nu}| + |f \lim \phi^* \beta - m.2^{-\nu}| < 2^{-\nu+1} .$$ Q.e.d.

13.3.2. If we adapt the proof of 13.1.4 we obtain a proof of the intuitionistic form of Heine-Borel for $[-1, 1]$. More generally, we can prove an intuitionistic form of Heine-Borel for every located compact space. See [H 1966] 5.2.2.

§ 14. Well-orderings and ordinals

14.1.

In this section we discuss a simplified version of Brouwers theory of well-orderings and ordinals ([Br 1926], [H 1959]). In Brouwers original definition, the type of a well-ordered species may become arbitrarily high, and is in fact itself characterized by an arbitrary well-ordering. Therefore we shall use a standard coding of well-orderings by means of species of natural numbers, in order to avoid such conceptual difficulties.

14.1.1. Definition. Let $X_0, \ldots, X_n \subseteq N$. We define operations $+, \Sigma, \cdot$:

$$X_0 + \ldots + X_n = \{\langle i\rangle * x : 0 \leq i \leq n \wedge x \in X_i\} \text{ ; as a special case}$$

we have

$$X_0 + = \{\langle 0\rangle * x : x \in X_0\} .$$

We write

$$\hat{x} * X \text{ or } \langle x\rangle * X \text{ for } \{\langle x\rangle * y : y \in X\} .$$

The infinite sum is defined by

$$X_0 + X_1 + \ldots = \Sigma_i X_i = \{\langle i\rangle * x : i \in N \wedge x \in X_i\} .$$

$$X_1 . X_0 = X_1 X_0 = \{n * m : n \in X_0 \wedge m \in X_1\} .$$

Now we introduce a class of species, WO, by a generalized inductive definition (analogous to the introduction of K is section 9).

14.1.2. Definition. Every $F \in WO$ (WO = the class (species) of well ordered species) is constructed by means of the following principles:

(a) $\{0\} \in WO$.

(b) $F_0, \ldots, F_n \in WO \rightarrow F_0 + \in WO \wedge F_0 + \ldots + F_n \in WO$.

(c) $\langle F_n\rangle_n$ lawlike, $\bigwedge n(F_n \in WO)$, then $\Sigma_i F_i \in WO$.

14.1.3. Convention. In this section, we reserve the capitals F, G, H, J for elements of WO.

Corresponding to the generalized inductive definition of WO we have the

14.1.4. Principle of proof by induction over WO.

If the following hypotheses (a), (b), (c) are satisfied

(a) $\{0\} \in X$,

(b) $F_0, \ldots, F_n \in X \to F_0 + \in X \wedge F_0 + \ldots + F_n \in X$,

(c) $\bigwedge n(F_n \in X)$, $\langle F_n \rangle_n$ lawlike $\to \underset{n}{\Sigma} F_n \in X$.

then

$$WO \subseteq X.$$

Just as in the case of ordinary induction, we may justify definition by induction over WO by the principle of proof by induction over WO.

14.1.5. Definition. $\prec$ is an ordering on the natural numbers defined as follows.
Let $n = \langle x_0, \ldots, x_r \rangle$, $m = \langle y_0, \ldots, y_s \rangle$.
$n \prec m \leftrightarrow \bigvee i(i \leq r \wedge i \leq s \wedge x_i < y_i \wedge \bigwedge j < i(x_j = y_j)) \vee (r > s \wedge \bigwedge j \leq s(x_j = y_j))$.

This ordering figures in the literature as the so-called Brouwer-Kleene ordering between finite sequences of natural numbers.

$\prec$ is a refinement of the partial ordering on sequences defined by
$n <^0 m \leftrightarrow \bigvee n'(n' \neq 0 \wedge n * n' = m)$.

$\prec$ is a total ordering, hence $\prec$ induces a total ordering on any $F \in WO$. Remark that $F_0 + \ldots + F_n$, $\underset{n}{\Sigma} F_n$ correspond to the formation of ordered sums in the usual sense. In the sequel, any terminology involving order on $F \in WO$ refers to the standard ordering $\prec | F$ (restriction of $\prec$ to F).

14.1.6. Definition. If X, Y are two (partially) ordered species, we write $X \sim Y$ to indicate the existence of a one-to-one mapping ϕ from X onto Y such that ϕ, ϕ^{-1} are order preserving. (In short: ϕ is an order isomorphism.)

The following properties of WO are easily proved by induction over WO:
14.1.7. Theorem.

(I) Every F is decidable.

(II) Every F possesses a first element.

(III) For all $x \in F$ $(F \in WO)$ either there is an immediate successor in F or x is a last element of F.

(IV) $\bigwedge F \bigwedge n \bigwedge m\ (n \in F \wedge n * m \in F \to m = 0)$. This property is more simply expressed as $\bigwedge F \bigwedge a \bigwedge x \bigwedge y\ (\bar{a}x \in F \wedge \bar{a}y \in F \to x = y)$.

The proofs are left as an exercise to the reader.

14.1.8. Theorem (Transfinite induction with respect to $\prec|F$).
For any species $X \subseteq F$:

$$\bigwedge x \in F[\bigwedge y \in F(y \prec x \to Xy) \to Xx] \to \bigwedge x \in F(Xx).$$

Proof. By induction over WO with respect to F. For $F = \{0\}$ the theorem is evident. Suppose now the assertion of the theorem to be valid for $F_1, F_2, F_3, \dots$. From this hypothesis we prove for any n (by ordinary induction)

$$\bigwedge x \in F_0+\dots+F_n[\bigwedge y \in F_0+\dots+F_n(y \prec x \to Xy) \to Xx] \to \bigwedge x \in F_0+\dots+F_n(Xx). \quad (1)$$

For $n = 0$, this is immediate. Suppose (1) to be proved up to $n = k$, and let

$$\bigwedge x \in F_0+\dots+F_{k+1}[\bigwedge y \in F_0+\dots+F_{k+1}(y \prec x \to Xy) \to Xx]. \quad (2)$$

This implies

$$\bigwedge x \in F_0+\dots+F_k[\bigwedge y \in F_0+\dots+F_k(y \prec x \to Xy) \to Xx]$$

and hence by our induction hypothesis

$$\bigwedge x \in F_0+\dots+F_k(Xx). \quad (3)$$

Now take an arbitrary $z \in \langle k+1 \rangle * F_{k+1}$.

As a consequence of (3) we have

$$\bigwedge y \in F_0+\dots+F_{k+1}(y \prec z \to Xy) \leftrightarrow \bigwedge y \in \langle k+1 \rangle * F_{k+1}(y \prec z \to Xz),$$

therefore from (2)

$$\bigwedge z \in \langle k+1 \rangle * F_{k+1}[\bigwedge y \in \langle k+1 \rangle * F_{k+1}(y \prec z \to Xy) \to Xz],$$

and this yields, since the theorem was supposed to hold for F_{k+1},

$$\bigwedge z \in \langle k+1 \rangle * F_{k+1}(Xz).$$

Hence, with (3)

$$\bigwedge x \in F_0+\dots+F_{k+1}(Xx).$$

Now if $F = \Sigma_i F_i$, then any $x \in F$ belongs to $F_0+\dots+F_k$ for some k, hence by the previous argument Xx.

14.1.9. Theorem. Let F be any well-ordered species. Every strictly descending (w.r.t $\prec$) sequence contains an element outside F.

Proof. Take Xx to be: every strictly descending sequence $\langle x_n \rangle_n$ with $x_0 = x$ contains an element outside F. Clearly

$$\bigwedge x \in F[\bigwedge y \in F(y \prec x \to Xy) \to Xx],$$

hence (by the previous theorem) $\bigwedge x \in F(Xx)$, i.e. every strictly descending sequence beginning with an element of F contains an element outside F. Since for an arbitrary strictly descending sequence $x_0 \notin F \vee x_0 \in F$, this proves our theorem.

14.2.

14.2.1. Theorem. $G(\sum_i F_i) = \sum_i GF_i$; $G(F_0+\ldots+F_n) = GF_0+\ldots+GF_n$.

Proof. immediate.

14.2.2. Theorem $\bigwedge F \bigwedge G(FG \in WO)$.

Proof. by induction on WO with respect to F, using the previous theorem.

14.2.3. Definition. For any $u, v \in F$ we define

$$F[u) = \{x : u \preceq x \wedge x \in F\},$$
$$F[u,v) = \{x : u \preceq x \prec v \wedge x \in F\},$$
$$F(v) = \{x : x \prec v \wedge x \in F\},$$

where $u \preceq v$ is an abbreviation for $u \prec v \vee u = v$.

$F(v)$, $F[u,v)$, $F[u)$ are detachable subspecies of F. For any v, $F(v)$ is empty or an element of WO.

For $u \prec v$, $F[u,v) = F[u) \cap F(v)$. $F = F(u) \cup F[u)$.

14.2.4. Definition ϕ_u is a mapping defined as follows.

Let $u = \langle x_0,\ldots, x_t, x_{t+1},\ldots, x_{t+s}\rangle = \langle 0,\ldots, x_{t+1},\ldots; x_{t+s}\rangle$.

(a) $v \prec u \to \phi_u v = v$

(b) $u \preceq \langle y_0,\ldots, y_t\rangle$, for some $i \leq t$ $y_i > 0 \to \phi_u\langle u_0,\ldots, y_r\rangle = \langle y_0,\ldots, y_r\rangle$.

(c) $u \preceq \langle y_0,\ldots, y_r\rangle$, $t < r \leq t+s$, $y_0 = \ldots = y_t = 0 \to$

$\phi_u\langle y_0,\ldots, y_r\rangle = \langle y_0,\ldots, y_t, y_{t+1} - x_{t+1},\ldots, y_r - x_r\rangle$.

(d) If $u \preceq \langle y_0,\ldots, y_r\rangle$, $t+s < r \to$

$\phi_u\langle y_0,\ldots, y_r\rangle = \langle y_0,\ldots, y_t, y_{t+1} - x_{t+1},\ldots, y_{t+s} - x_{t+s}, y_{t+s+1},\ldots, y_r\rangle$.

It may be helpful to the reader to verify by representing a simple finite F by a construction tree (see the picture below) what exactly ϕ_u does.

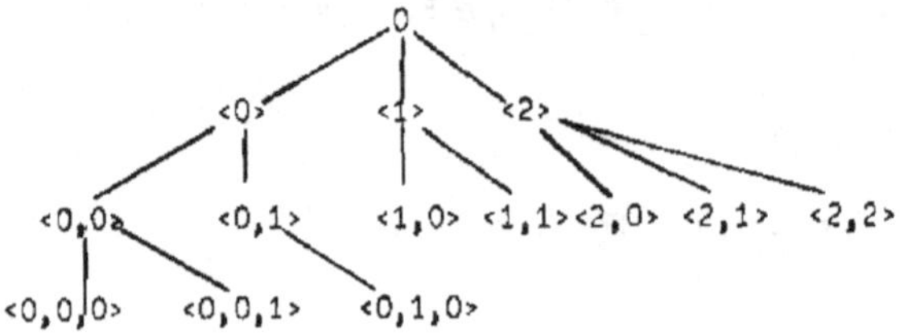

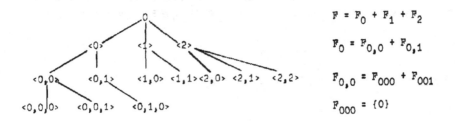

$F = F_0 + F_1 + F_2$

$F_0 = F_{0,0} + F_{0,1}$

$F_{0,0} = F_{000} + F_{001}$

$F_{000} = \{0\}$

(The terminal nodes of the tree represent the elements of F.)

We use in the remainder of this section S,T as letters for species of WO and species $F[u)$, $F(u)$, $F[u,v)(u,v \in F,\ F \in \mathrm{WO})$.

We introduce a mapping Φ from species into species; Φ is defined for any S. If u is the first element of S, then $\Phi S = \{\phi_u x : x \in S\}$.

Clearly $\bigwedge S(\Phi S \in \mathrm{WO}))$by induction), and $\bigwedge S(S \sim \Phi S)$.

14.2.5. <u>Theorem</u>. Let F be an arbitrary element of WO, and let $\langle n_i \rangle_i$ be a strictly increasing sequence of elements of F. Then

$F = F(n_1) \cup F[n_1,n_2) \cup F[n_2,n_3) \ldots \sim F(n_1) + \Phi F[n_1,n_2) + \Phi F[n_2,n_3) + \ldots$

<u>Proof</u>. Since $n_0 < n_1, n_1$ is not the first element of F, hence $F(n_1)$ is inhabited. Otherwise the proof is routine.

14.2.6. <u>Definition</u>. $F < G \equiv_D \bigvee x(x \in G \wedge F \sim G(x))$.
(Equivalently: $F < G \equiv_D \bigvee H(F+H \sim G)$.

14.2.7. <u>Theorem</u>.
(I) If ϕ is an order-isomorphism between F and a subset $X \subseteq F$, then $X = F$ and ϕ is the identity.

(II) $\bigwedge F \bigwedge G \bigwedge x,\ y \in G(F \sim G(x) \wedge F \sim G(y) \rightarrow x = y)$.

<u>Proof</u>. (I). It is possible to prove

$$\bigwedge x \in F(\bigwedge y \in F(y \prec x \rightarrow \phi y = y) \rightarrow \phi x = x).$$

For suppose $\phi x \prec x \wedge \bigwedge y \in F(y \prec x \to \phi y = y)$; then apparently $\phi\phi x = \phi x$, hence $\phi^{-1}\phi\phi x = \phi^{-1}\phi x$, i.e. $\phi x = x$. Therefore $\phi x = x$ or $\phi x \succ x$. Suppose $x \prec \phi x$. Then $\phi^{-n-1}x \prec \phi^{-n}x$ for every n, and so $\langle \phi^{-n}x \rangle_n$ would be an infinite descending sequence; and this conflicts with 14.1.9. Therefore $\phi x = x$. Now we apply 14.1.8 and obtain $\bigwedge x \in F(\phi x = x)$.

(II). Let $x \prec y$, $F \sim G(x)$, $F \sim G(y)$, then from $F \sim G(x)$, $F \sim G(y)$ we conclude to the existence of an order isomorphism from $G(y)$ onto $G(x)$, in contradiction with (I). Likewise for $x \succ y$; hence $x = y$.

14.2.8. Definition. For any $F < G$ we define: if $F \sim G(u)$, then $G-F \equiv_D \phi G[u)$. From the preceding theorem it follows that this definition is unique.

14.2.9. Theorem.

(I) $F \sim F', G \sim G' \to F.G \sim F'.G'$.

(II) $F < G \to F + (G - F) \sim G$.

(III) $F < G \to H.F < H.G$.

(IV) $F < G \to H.(G - F) \sim H.G - H.F$.

Proof. (I) is evident, (II) follows from the definitions, (III) and (IV) are proved simultaneously as follows.

$$H.F + H.(G-F) \sim H.(F + (G-F)) \sim H.G \quad .$$

Hence $HF < HG$, $H(G-F) \sim HG-HF$ (with 14.2.8 II).

14.2.10. Definition. Let $\langle H_n \rangle_n$ be a lawlike sequence such that $\bigwedge i(H_i < H_{i+1})$. Then $\lim_n H_n = \Sigma_n H'_n$ with $H'_0 = H_0$, $H'_n = H_n - H_{n-1}$ for $n > 1$.

14.2.11. Lemma.

(I) $\bigwedge n(H_n \sim J_n) \to \lim_n H_n \sim \lim_n J_n$.

(II) $G . \lim_n H_n \sim \lim_n G . H_n$.

Proof. (I) is immediate.

(II) $G . \lim_n H_n = G(H_0 + (H_1 - H_0) + (H_2 - H_1) + \ldots) =$

$G . H_0 + G.(H_1 - H_0) + \ldots \sim GH_0 + (GH_1 - GH_0) + (GH_2 - GH_1) + \ldots = \lim_n GH_n$.

In view of the next definition we remark that it is decidable whether an F contains exactly one or more than one element.

14.2.12. Definition. For any $F, G \in WO$ we define $F^G = F \exp G$ inductively as follows.

For $F \sim \{0\}$ we put $F^G = F$ for any G. For $F > \{0\}$ we put

(a) $F^{\{0\}} = F$.

(b) $F \exp (G_0 + \ldots + G_n) = F^{G_0} F^{G_1} \ldots F^{G_n}$.

(c) $F \exp \sum_i G_i = \lim_n F^{G_0} \ldots F^{G_n}$.

14.2.13. Lemma. Let $\langle k(n) \rangle_n$ be a strictly increasing (in the natural ordering) lawlike sequence of natural numbers, and let $\langle H_n \rangle_n$ be a lawlike sequence such that $\bigwedge n(H_n < H_{n+1})$. Then

$$\lim_n H_n \sim \lim_n H_{k(n)} .$$

Proof. We remark that if $F_1 < F_2 < F_3$, then $F_3 - F_1 \sim (F_2 - F_1) + (F_3 - F_2)$, since $F_1 + ((F_2 - F_1) + (F_3 - F_2)) \sim (F_1 + (F_2 - F_1)) + (F_3 - F_2) \sim F_2 + (F_3 - F_2) \sim F_3$

Repeated application of this result yields:

$$H_{k(n+1)} - H_{k(n)} \sim (H_{k(n)+1} - H_{k(n)}) + (H_{k(n)+2} - H_{k(n)+1}) + \ldots$$
$$\ldots + (H_{k(n+1)} - H_{k(n+1)-1}) . \qquad (4)$$

One verifies easily that from $\bigwedge i(G_i \sim J_{w(i)} + \ldots + J_{w(i+1)-1})$, $w(0) = 0$, one obtains $\sum_i G_i \sim \sum_i J_i$. Hence by (4) we conclude to $\lim_n H_n \sim \lim_n H_{k(n)}$.

14.2.14. Lemma. Let $\langle k(n) \rangle_n$ be a strictly increasing (in the natural order) sequence of natural numbers, and let $\langle H_n \rangle_n$ be a lawlike sequence.

(I) If $\bigwedge n(H_n < H_{n+1})$, $F > \{0\}$, then

$$\lim_n F \exp (H_0 + \ldots + H_n) \sim \lim_n F \exp (H_0 + \ldots + H_{k(n)}).$$

(II) Let $\bigwedge i(H_i = G_{k(i)} + \ldots + G_{k(i+1)-1})$, $k(0) = 0$. Then

$$F \exp \sum_i G_i \sim F \exp \sum_i H_i .$$

(III) Let $\bigwedge i < n(H_i = G_{k(i)} + \ldots + G_{k(i+1)-1})$, $k(0) = 0$, and $H_n = \sum_j G_{k(n)+j}$. Then $F \exp \sum_i G_i \sim F \exp(H_0 + \ldots + H_n)$.

Proof.

(I) For $F > \{0\}$, $F^G > \{0\}$. Hence $\bigwedge i(F \exp (H_0 + \ldots + H_{i+1}) > F \exp (H_0 + \ldots + H_i)$. Then apply 14.2.13.

(II) $F \exp \sum_i H_i = \lim_i F \exp (H_0 + \ldots + H_i) = \lim_i F \exp(G_0 + \ldots + G_{k(n+1)-1}) \sim \lim_i F \exp (G_0 + \ldots + G_i)$ (by (I)) $\sim F \exp \sum_n G_n$.

(III) $F \exp (H_0 + \ldots + H_n) = F \exp H_0 \cdot \ldots \cdot F \exp H_n =$

$F \exp (G_0 + \ldots + G_{k(n)-1}) \; F \exp \sum_j G_{k(n)+j} =$

$F \exp (G_0 + \ldots + G_{k(n)-1}) \lim_j F \exp (G_{k(n)} + \ldots + G_{k(n)+j}) \sim$

$\lim_j F \exp (G_0 + \ldots + G_{k(n)+j})$ (14.2.11, II) $\sim \lim_j F \exp (G_0 + \ldots + G_j)$

$= F \exp \sum_j G_j$ (by (I)).

14.2.15. Definition. H is called a refinement of F (notation $F \ast H$) if $H = H_0 + \ldots + H_n$ ($H = \sum_i H_i$), $H_i = \Phi S_i$, $i = 0, \ldots, n$ ($i \in N$ resp.), $F = S_0 \cup \ldots \cup S_n$ ($F = \bigcup_n S_n$ resp.), $\bigwedge i \bigwedge j \bigwedge x \in S_i \bigwedge y \in S_j$ $(i < j \rightarrow x \prec y)$.

14.2.16. Remark. The relation $\ast$ is evidently not symmetrical.

14.2.17. Lemma. When $G \ast H$, then $F^G \sim F^H$.

Proof. We prove this lemma by induction over WO with respect to G. We distinguish four cases in the induction step.

Case I. Let $G = G_0 + \ldots + G_n$, $G \ast H_0 + \ldots + H_m$.

Let $H_i = \Phi S_i$ for $i \leq n$, $S_0 \cup \ldots \cup S_n = F$. $\bigwedge i \bigwedge j \bigwedge x \in S_i \bigwedge y \in S_j$ $(i<j \rightarrow x \prec y)$.

Now we can find $T_0, \ldots, T_k$ such that

$$S_i = \langle x_i \rangle \ast T_{a(i)} \cup \ldots \cup \langle x_i + y_i \rangle \ast T_{a(i+1)-1}, \; i \leq n,$$
$$a(0) = 0, \; y_i = a(i+1) - a(i) - 1, \; \bigwedge i(a(i) < a(i+1)).$$
$$G_i = T_{b(i)} \cup \ldots \cup T_{b(i+1)-1}, \; i \leq m, \; b(0) = 0, \; \bigwedge i(b(i) < b(i+1)).$$

Take $\Phi T_i = J_i$ for $0 \leq i \leq k$. Then $H_i = \Phi S_i =$

$\Phi(\langle x_i \rangle \ast T_{a(i)} \cup \langle x_i + 1 \rangle \ast T_{a(i)+1} \cup \ldots) = \langle 0 \rangle \ast \Phi T_{a(i)} \cup$

$\cup \langle 1 \rangle \ast \Phi T_{a(i)+1} \ldots = \Phi T_{a(i)} + \ldots + \Phi T_{a(i+1)-1} = J_{a(i)} + \ldots + J_{a(i+1)-1}$.

Also $G_i \ast J_{b(i)} + \ldots + J_{b(i+1)-1}$, $i \leq m$, $b(0) = 0$.

Now, using our induction hypothesis with respect to $G_0, \dots, G_n$:

$$F^G = F^{G_0} \cdot \dots \cdot F^{G_n} \sim F^{J_0} \cdot \dots \cdot F^{J_k} \sim F^{H_0} \cdot \dots \cdot F^{H_m} = F^H.$$

Case II. $G = G_0 + \dots + G_n$, $G \approx \sum_i H_i$. By a reasoning analogous to the argument in the previous case, we can find J_i, $i \in N$ such that

$$G_i \approx J_{a(0)} + \dots + J_{a(i+1)-1}, \; i < n, \; a(0) = 0, \; \Lambda i(a(i) < a(i+1)),$$
$$G_n \approx \sum_k J_{a(n)+k}$$
$$H_i = J_{b(i)} + \dots + J_{b(i+1)-1}, \; i \in N, \; b(0) = 0, \; \Lambda i(b(i) < b(i+1)).$$

Then, with the use of the induction hypothesis:

$$F^G = F^{G_0} \dots F^{G_n} \sim F \exp J_{a(0)} \dots F \exp J_{a(n)-1} \, F \exp \sum_k J_{a(n)+k} \sim$$
$$F \exp (J_{a(0)} + \dots + J_{a(n)-1}) \lim_k F \exp (J_{a(n)} + \dots + J_{a(n)+k}) \sim$$
$$\lim_k F \exp (J_{a(0)} + \dots + J_{a(n)+k}) \sim \lim F \exp (J_0 + \dots + J_k)$$

(14.2.11 (II), 14.2.14).

Case III. $G = \sum_i G_i$, $G \approx H_0 + \dots + H_n = H$. Now we can find J_i, $i \in N$, such that

$$G_i \approx J_{a(i)} + \dots + J_{a(i+1)-1}, \; i \in N, \; a(0) = 0, \; \Lambda i(a(i) < a(i+1))$$
$$H_i = J_{b(i)} + \dots + J_{b(i+1)-1}, \; i < n, \; b(0) = 0, \; \Lambda i(b(i) < b(i+1))$$
$$H_n = \sum_k J_{b(n)+k} .$$

Then $F^G = F^H$ is proved from the induction hypothesis in the same manner as in the previous case.

Case IV. $G = \sum_i G_i$, $G \approx \sum_i H_i = H$. Now we can find J_i, $i \in N$, such that

$$G_i \approx J_{a(i)} + \dots + J_{a(i+1)-1}, \; a(0) = 0, \; i \in N, \; \Lambda i(a(i) < a(i+1))$$
$$H_i = J_{b(i)} + \dots + J_{b(i+1)-1}, \; b(0) = 0, \; i \in N, \; \Lambda i(b(i) < b(i+1)),$$

etc.etc.

14.2.18. Theorem. $G \sim H \rightarrow F^G \sim F^H$.

Proof. By induction over WO with respect to G. Let $G = G_0 + \dots + G_n$. Then to every G_i we can find a J_i, $J_0 = \Phi H(v_0)$, $J_i = \Phi H[v_{i-1}, v_i)$ for $i < n$, $J_n = \Phi H[v_n)$, $J_i \sim G_i$ for $i \leq n$.

So $F^G = F^{G_0} \ldots F^{G_n} \sim F^{J_0} \ldots F^{J_n} \sim F^H$ (by the previous lemma, since $H = J_0 + \ldots + J_n$).

If $G = \sum_i G_i$, we can find J_i, $i \in N$, with $J_i \sim G_i$ for every i, $J_0 = \phi H(v_0)$, $J_i = \phi H[v_{i-1}, v_i)$ and $H = \sum_i J_i$. Therefore $F^G =$ $\lim_n F^{G_0 + \ldots + G_n} = \lim_n F^{J_0 + \ldots + J_n} = F \exp \sum_i J_i \sim F \exp H$ (by the previous lemma).

14.2.19. Theorem. $G \sim H \rightarrow G^F \sim H^F$.

Proof. by induction over WO with respect to F.

14.2.20. Theorem. $F \exp GH \sim (F \exp G) \exp H$.

Proof. We apply induction over WO with respect to H.

$F \exp G.\{0\} = F \exp G = (F \exp G) \exp \{0\}$. Now let $H = H_0 + \ldots + H_n$. Then

$F \exp G(H_0 + \ldots + H_n) = F \exp (GH_0 + \ldots + GH_n) \sim$

$F \exp GH_0 \cdot \ldots \cdot F \exp GH_n \sim (F \exp G)^{H_0} \ldots (F \exp G)^{H_n} \sim (F \exp G) \exp H.$

Let $H = \sum_i H_i$. Then $F \exp G(\sum_i H_i) = F \exp \sum_i GH_i =$
$\lim_n F \exp (GH_0 + \ldots + GH_n) \sim \lim_n (F \exp G) \exp (H_0 + \ldots + H_n) \sim$
$\lim_n (F \exp G)^{H_0} \ldots (F \exp G)^{H_n} = (F \exp G) \exp H.$

14.3.

14.3.1. Definition. We define an ordered sum of a species $\{F_x : x \in G\}$ with $\bigwedge x \in G\ (F_x \in WO)$ by induction over WO with respect to G:

(a) $G = \{0\} \rightarrow \Sigma\{F_x : x \in G\} = F_0 +$

(b) $G = G_0 + \ldots + G_n \rightarrow \Sigma\{F_x : x \in G\} =$

$$\Sigma\{F_{\langle 0\rangle * y} : y \in G_0\} + \ldots + \Sigma\{F_{\langle n\rangle * y} : y \in G_n\}.$$

(c) $G = \sum_i G_i \rightarrow \Sigma\{F_x : x \in G\} = \sum_z (\Sigma\{F_{\langle z\rangle * y} : y \in G_z\})$.

14.3.2. Lemma. Let $F \simeq G$, and let ϕ be an order-isomorphism from F onto H. Suppose $\bigwedge x \in F\ (H_x = H'_{\phi x})$, then

$$\Sigma\{H_x : x \in F\} \sim \Sigma\{H'_{\phi x} : x \in G\}.$$

Proof. As in 14.2.17, we apply induction over WO with respect to F, and we have to distinguish four cases. Let us treat as a typical case $F = \sum_i F_i$, $G = \sum_i G_i$. Then we can find J_i such that

$$F_i = J_{a(i)} +\ldots+ J_{a(i+1)-1},\ a(0) = 0,\ i \in N,\ \Lambda i(a(i) < a(i+1)),$$

$$G_i = J_{b(i)} +\ldots+ J_{b(i+1)-1},\ b(0) = 0,\ i \in N,\ \Lambda i(b(i) < b(i+1)).$$

$$\Sigma\{H_x : x \in F\} = \sum_z \Sigma\{H_{<z>* y} : y \in F_z\} \sim \sum_z \Sigma\{\Sigma\{\bar{H}_{z,y} : y \in J_{a(z)} +\ldots+ J_{a(z+1)-1}\}\},$$

where $\bar{H}_{z,y}$ is such that if $\lambda y.\ \psi(z, y)$ is an order-isomorphism between F_z and $J_{a(z)} +\ldots+ J_{a(z+1)-1}$ for every z, then $\Lambda z\, \Lambda y(\bar{H}_{z,\psi(z,y)} = H_{<z>* y})$.
The conclusion then follows from the induction hypothesis.

Hence $\Sigma\{H_x : x \in F\} \sim \sum_z \Sigma\{\Sigma\{\hat{H}_{z,y} : y \in J_z\}\}$, with $\hat{H}_{z,y}$ such that

$\Lambda z \Lambda v \Lambda y\ (0 \leq v \leq a(z+1)-a(z) -1 \wedge y \in J_{a(z)+v} \rightarrow \bar{H}_{z,<v>* y} = \hat{H}_{a(z)+v,y})$.

$\sum_z\{\Sigma\{\hat{H}_{z,y} : y \in J_z\}\} \sim \sum_z\{\Sigma\{\hat{H}_{a(z),y} : y \in J_{a(z)}\} +\ldots+ \Sigma\{\hat{H}_{a(z+1)-1,\,y} : y \in J_{a(z+1)-1}\}\} =$

$\sum_z \Sigma\{H'_{<z>* y} : y \in G_z\} = \Sigma\{H'_y : y \in G\}$, if we take H'_y to be such that

$\Lambda z \Lambda u \Lambda y\ (0 \leq u \leq b(z+1)-b(z)-1 \wedge y \in J_{b(z)+u} \rightarrow H'_{<z,u>* y} = \hat{H}_{b(z)+u,y})$.

14.3.3. Theorem. Let $G \sim H$, and let ϕ be an order-isomorphism between G and H; suppose $\Lambda x \in G\ (F_x = J_{\phi x})$; then $\Sigma\{F_x : x \in G\} = \Sigma\{J_{\phi x} : x \in H\}$.
Proof. Completely analogous to the derivation of 14.2.18 from 14.2.17.

14.3.4. Theorem. Let $\{F_x : x \in G\}$ and $\{H_x : x \in G\}$ be given such that $\Lambda x \in G\ (F_x \sim H_x)$, then $\Sigma\{F_x : x \in G\} = \Sigma\{H_x : x \in G\}$.
Proof. by induction over WO with respect to G.

14.4.

14.4.1. Definition. Ordinals are the equivalence classes of order-isomorphic species of WO. The ordinal $\alpha = \text{Ord}\ (F)$ associated with a species $F \in WO$ is therefore defined as $\{G : G \sim F\}$.

$\text{Ord}\ (\{0\}) = 1$, $\text{Ord}\ (\{<i> : i \in N\}) = \omega$.

The theorems of 14.2 permit us to define a few arithmetical operations on ordinals.

14.4.2. <u>Definition.</u> $Ord(F + \{0\}) = Ord(F) + 1.$
$Ord(F + G) = Ord(F) + Ord(G).$
$Ord(F^G) = Ord(F)^{Ord(G)} = Ord(F) \exp Ord(G).$
$Ord(FG) = Ord(F).\ Ord(G).$

Some simple identities of ordinal arithmetic follow readily from the theorems proved for well-ordered species:

$$(\alpha + \beta) + \gamma = \alpha + (\beta + \gamma).$$
$$(\alpha\beta)\gamma = \alpha(\beta\gamma).$$
$$\alpha(\beta + \gamma) = \alpha\beta + \alpha\gamma.$$
$$\alpha^{\beta+\gamma} = \alpha^\beta\alpha^\gamma.$$
$$(\alpha^\beta)^\gamma = \alpha^{\beta\gamma}.$$

14.4.3. If we define $\alpha < \beta$ by $V\gamma(\alpha + \gamma = \beta)$ (recall that we have excluded 0 as an ordinal for reasons of technical convenience) then it may be concluded from 14.3 that we can define well ordered sums of ordinals unambiguously: $\Sigma\{\alpha_\beta : \beta < \gamma\}$.

14.4.4. It is a consequence of 14.1.5, 14.2.7 (II), that $\{\alpha : \alpha < \beta\}$ is totally ordered for any given β. But on the other hand we have no hope of proving the species of all ordinals to be totally ordered, as is illustrated by the following (weak) counterexample.

Let a be an arbitrary lawlike function of $(N)N$.
Define the sequence $\langle F_n \rangle_n$: $an = 0 \rightarrow F_n = \{0\}$, $an \neq 0 \rightarrow F_n = \{\langle i \rangle : i \in N\}$. Then we have no general method to decide

$$Ord(\Sigma_n F_n) = \omega \vee \omega < Ord(\Sigma_n F_n)$$

since this would imply

$$\Lambda a(\Lambda n(an = 0) \vee Vn\ (an \neq 0)).$$

For a concrete illustration we only have to make a depend on the decimals of π in the usual way.

We shall refrain here from a further development of ordinal arithmetic; the reader is referred to [Br 1926].

14.5.

14.5.1. It will be clear that there is much similarity in behaviour between K and WO. By a quite general theorem one obtains immediately that WO is explicitly definable in terms of K. (For a statement of the theorem, see section 15.)

Here we shall demonstrate by a more straightforward argument that K and WO are explicitly definable in terms of each other, and that the closure conditions and induction scheme for K, WO respectively are provable from these explicit definitions.

For an easy comparison, we study W, the species of characteristic functions of elements of WO instead of WO itself.

If we put

$$A_W(P,a) \equiv_D [a = \lambda x.(1 \dot{-} x)] \vee [a0 = 0 \wedge \bigwedge y\ \lambda x.a(\hat{y} * x) \in P] \vee$$
$$\vee [a0 = 0 \wedge \bigvee z \bigwedge y((y \leq z \wedge \lambda x.a(\hat{y} * x) \in P) \vee (y > z \wedge \bigwedge x\, a(\hat{y} * x) = 0))]$$

then the introduction of W by a generalized inductive definition is expressed by

$$A_W(W,a) \rightarrow Wa,$$
$$\bigwedge a[A_W(Q,a) \rightarrow Qa] \rightarrow [W \subseteq Q].$$

14.5.2. <u>Theorem</u>. K is explicitly definable in terms of W such that

$$A_K(K,a) \rightarrow Ka,$$
$$\bigwedge a[A_K(Q,a) \rightarrow Qa] \rightarrow [K \subseteq Q]$$

become provable (by induction over W).

<u>Proof</u>. If we replace $a = \lambda x.(1 \dot{-} x)$ in the definition of $A_W(Q,a)$ by $\bigvee z[a = \lambda x.(z+1)]$, we obtain $A_{W^*}(Q,a)$.

If we define a species W^* by

$$a \in W^* \leftrightarrow \bigvee b \in W \bigvee c(a = \lambda n.bn.\, sg\, h(b,n).(ch(b,n) + 1))$$

where $sg\, x = 1 \dot{-} (1 \dot{-} x)$, $h(b,n) = m + 1$ if $bm \neq 0$, $m \leq^0 n$, $m'(m' <^0 m \rightarrow bm' = 0)$, $h(b,n) = 0$ otherwise.

Then one proves easily

$$\bigwedge a(A_{W^*}(W^*,a) \rightarrow W^*a),$$
$$\bigwedge a[A_{W^*}(Q,a) \rightarrow Qa] \rightarrow [W^* \subseteq Q].$$

The proof is left to the reader.

Now K is explicitly definable by

$$a \in K \leftrightarrow a \in W^* \wedge \bigwedge n \bigvee y\ a(n * y) \neq 0.$$

$$A_K(K,a) \rightarrow Ka$$

is readily proved. Suppose now

$$\bigwedge a[A_K(Q,a) \rightarrow Qa]. \qquad (4)$$

Take $Q^0 a \equiv_D \bigwedge n \bigvee y\ a(n * y) \neq 0 \rightarrow Qa.$

If $a = \lambda x.z+1$, then $A_K(Q,a)$, hence Qa.
Now let

$$a0 = 0 \wedge \bigwedge y(\bigwedge n \bigvee z\ a(\hat{y} * n * z) \neq 0 \rightarrow \lambda m.a(\hat{y} * m) \in Q),$$

then

$$a0 = 0 \wedge \bigwedge n \bigvee z(a(n * z) \neq 0) \rightarrow \bigwedge y\ \lambda m.a(\hat{y} * m) \in Q,$$

hence

$$\bigwedge n \bigvee z(a(n * z) \neq 0) \rightarrow (a0 = 0 \wedge \bigwedge y\ \lambda m.a(\hat{y} * m) \in Q),$$

therefore by (4)

$$\bigwedge n \bigvee z\ a(n * z) \neq 0 \rightarrow a \in Q, \text{ so } a \in Q^0.$$

If

$$a0 = 0 \wedge \bigvee z \bigwedge y[(y \leq z \wedge \lambda x.a(\hat{y} * x) \in Q^0) \vee (y > z \wedge \bigwedge x\ a(\hat{y} * x) = 0)]$$

we have trivially $Q^0 a$, since $\bigwedge n \bigvee y\ a(n * y) \neq 0$ clearly does not hold.

Hence we have shown that

$$(4) \rightarrow \bigwedge a[A_{W^*}(Q^0,a) \rightarrow Q^0 a]$$

and therefore $W^* \subseteq Q^0$, i.e. $\bigwedge a(a \in W^* \rightarrow a \in Q^0)$.
Hence

$$\bigwedge a(a \in W^* \wedge \bigwedge n \bigvee y\ a(n * y) \neq 0 \rightarrow a \in Q^0 \wedge \bigwedge n \bigvee y\ a(n * y) \neq 0)$$

so $K \subseteq Q$.

14.5.3. <u>Theorem</u>. W is explicitly definable in terms of K, such that

$$\bigwedge a[A_W(W,a) \rightarrow Wa]$$

$$\bigwedge a[A_W(Q,a) \rightarrow Qa] \rightarrow [W \subseteq Q]$$

becomes provable (by induction over K).

<u>Proof</u>. Let V denote a class of spread directions satisfying

$$Va \equiv_D a0 \neq 0 \wedge \Lambda n(an \neq 0 \to [\Lambda x\ a(n * \hat{x}) \neq 0 \vee$$
$$\vee \mathsf{V}z\{\Lambda x \leq z(a(n * \hat{x}) \neq 0) \wedge \Lambda x > z(a(n * \hat{x}) = 0)\}]).$$

We define W^*:

$b \in W^* \equiv_D \mathsf{V}a \in V\ \mathsf{V}e\ \Lambda n(bn \neq 0 \leftrightarrow \mathsf{V}m(m \leq^0 n \wedge sg(am).em = bn \neq 0))$.

We prove easily

$$A_{W^*}(W^*, a) \to W^*a.$$

Now suppose

$$\Lambda a[A_{W^*}(Q,a) \to Qa]. \tag{5}$$

We apply induction with respect to e to prove $W^* \subseteq Q$.
Let us introduce $\psi(e,a)$ $(e \in K,\ a \in V \vee a = \lambda x.0)$ for the unique function b such that

$$\Lambda n(bn \neq 0 \leftrightarrow \mathsf{V}m(m \leq^0 n \wedge bn = sg(am).em \neq 0)).$$

Now we want to prove by induction over K

$$\Lambda e \Lambda a \in V\ (\psi(e,a) \in Q).$$

If $e = \lambda x.z+1$, then $\psi(e,a) = \lambda x.z+1$, hence $\psi(e,a) \in Q$.
Let now

$$e0 = 0 \wedge \Lambda x \Lambda c \in V\ (\psi(\lambda n.e(\hat{x} * n), c) \in Q) \tag{6}$$

and let $a \in V$. We remark that

$$a \in V \to \Lambda x(\lambda n.a(\hat{x} * n) \in V) \vee$$
$$\mathsf{V}z(\Lambda x \leq z(\lambda n.a(\hat{x} * n) \in V) \wedge \Lambda x > z(\Lambda n a(\hat{x} * n) = 0)).$$

Furthermore

$$a\hat{x} \neq 0 \wedge e0 = 0 \to \lambda n.\psi(e,a)(\hat{x} * n) = \psi(\lambda n.e(\hat{x} * n),\ a(\hat{x} * n)).$$

Suppose first $\Lambda x(\lambda n.a(\hat{x} * n) \in V)$. Then from (6):

$$\left.\begin{array}{l}\Lambda x(\lambda n.\psi(e,a)(\hat{x} * n) \in Q \\ \psi(e,a) = 0\end{array}\right\} \tag{7}$$

and thus from (5) and (7) (which implies $A_{W^*}(Q,\psi(e,a))$) we conclude to $Q\psi(e,a)$.

Now let $\mathsf{V}z(\Lambda x \leq z(\lambda n.a(\hat{x} * n) \in V) \wedge \Lambda x > z(\Lambda n\ a(\hat{x} * n) = 0))$.
Then from (6):

$$\mathsf{V}z(\Lambda x \leq z(\lambda n.\psi(e,a)(\hat{x} * n) \in Q) \wedge \Lambda x >z\ \Lambda n(\psi(e,a)(\hat{x} * n) = 0))$$

and

$$\psi(e,a) = 0,$$

therefore $A_{W^*}(Q,\psi(e,a))$, hence with (5) $Q\psi(e,a)$.
Thus we have proved $\Lambda e \Lambda a \in V(\psi(e,a) \in Q)$, i.e. $W^* \subseteq Q$.

§ 15. Species revisited; the role of the comprehension principle

15.1.

In the intuitionistic theory of species, we are confronted with the analogue of the fundamental question of classical axiomatic set theory: which species may be said to exist? Or in terms of a theory of constructions: which notions are constructions? The problem is illustrated by the role of the comprehension principle.

Let X be a certain given species. Once X is accepted as a well-defined object, there is nothing problematic in accepting X^n for any $n \in N$.

Let $\mathcal{L}$ be a language containing variables x, y for elements of X. The comprehension principle relative to $\mathcal{L}$, X may be expressed by a schema:

For every formula $F(x)$ of $\mathcal{L}$ not containing y we accept the universal closure of

$$\bigvee Y \bigwedge x\ (Yx \leftrightarrow F(x)) \qquad (1)$$

(Y a variable for subspecies of x.)

The strength of a comprehension principle relative to X clearly depends on the expressive power of $\mathcal{L}$. When $\mathcal{L}$ is a first order language with variables for elements of X, the resulting predicative version of the comprehension principle seems to us to be quite unobjectionable. Let us call this weak comprehension. Evidently we can build a ramified hierarchy by repeated weak comprehension, as in classical ramified analysis.

Strong or full comprehension, where $\mathcal{L}$ contains quantifiers $\bigwedge Y$, $\bigvee Y$, represents the other extreme possibility.
The strength of full comprehension is illustrated by
15.1.1. Theorem (see e.g. [Kr 1968 A]). Suppose $\mathcal{L}$ to consist of intuitionistic arithmetic with the language extended by variables $X, Y, \ldots$ for subspecies of N, and quantifiers $\bigwedge X, \bigvee X$, and the comprehension axiom relative to this language and N.
If $\mathcal{L}^+$ denotes the corresponding classical system, then the Gödel translation of § 3 extends to $\mathcal{L}^+$ (taking $(\bigwedge X\ F(X))^- = \bigwedge X\ F^-(X)$, $(\bigvee X\ F(X))^- = \neg\bigwedge X \neg F^-(X)$, and preserves validity.

Proof. We only have to verify that the translation of the comprehension scheme (1) for F is a consequence of the comprehension scheme in $\mathcal{L}$ applied to F^- and this is straightforward.
Therefore $\mathcal{L}^+$ is consistent if $\mathcal{L}$ is consistent.

15.2.

Inductive definitions like those of K in § 9, and the definition of WO in § 14 represent examples of intermediate forms of comprehension.

If we accept full comprehension, then the introduction of a predicate P_A satisfying

$$\left.\begin{array}{l} \Lambda x(A(P_A, x) \rightarrow P_A x) \\ \Lambda x(A(Q, x) \rightarrow Qx) \rightarrow P_A \subseteq Q \end{array}\right\} \qquad (2)$$

is justified (classically as well as intuitionistically) whenever A satisfies the condition of monotonicity:

$$A(P, x) \wedge P \subseteq P' \rightarrow A(P', x). \qquad (3)$$

P_A is said to be introduced by a generalized inductive definition (g.i.d). The justification is given by remarking that P_A may be defined by

$$P_A y \leftrightarrow \Lambda X\left[\Lambda x\ (A(X, x) \rightarrow Xx) \rightarrow Xy\right]. \qquad (4)$$

Since (4) requires universal quantification over species, we might, if we think this justification satisfactory, just as well accept full comprehension outright.

But if we impose more stringent requirements on A, sometimes better justifications for the introduction of P_A can be given. For example, the introduction of K (introduced and discussed in section 9) is essentially justified by observing that $A_K(P, a) \rightarrow Pa$ expresses closure of P under certain simple operations; and K is then viewed as the species such that $e \in K$ iff this can be proved using these simple closure properties only. Moreover, it is to be remarked that any $e \in K$ may be said (in a sense) to codify itself a standard proof of $e \in K$.

Likewise we may justify the introduction of WO. In fact, we have even shown WO to be explicitly definable in terms of K.

Once we have accepted K, a quite general class of g.i.d.'s also becomes acceptable, since the species P_A required to exist by the g.i.d. may also be defined explicitly in terms of K. We have the following result:

15.1.2. Theorem. Let IDK^* be an extension of the system IDK as described in § 10, and let A(P, a) be any formula of IDK^* with a single predicate letter P, such that A is constructed by means of $\wedge, \vee, \Lambda x, V x, V a$ from formulae not containing P and formulae of the form

$$P(\lambda y.t\ [b_1, b_2, \ldots, x_1, x_2, \ldots y]),\ t \text{ a term of } IDK^*.$$

Then we can explicitly define (in IDK^*) a predicate P_A such that

$$\Lambda a(A(P_A, a) \to P_A a),$$
$$\Lambda a[A(Q, a) \to Qa] \to \Lambda a[P_A a \to Qa]$$

for any Q in the language of IDK^*.

<u>Proof</u>. We shall not present a full proof here; for more details see [Kr, T]. The essential idea is that for $A(P, a)$ of the form described above, $a \in P_A$ must have a standard "cut-free" proof, which may be codified by a well-founded tree, hence by a function of K.

Let for example $A(P, a)$ be of the form

$$\vee b \Lambda x(R(a, b, x) \vee P\phi[a, b, x])$$

($\phi[a, b, x] = \lambda y.t\,[a, b, x, y]$ for a suitable term t).
Then we take as our explicit definition:

$$P_A a \equiv_D \vee e \vee c \vee d\ \{d^0 = a \wedge e0 = 0 \wedge \Lambda m \Lambda y\ ((e(m * \hat{y}) = 0 \to$$
$$\to d^{m * \langle y \rangle} = \phi[d^m, c^m, y]) \wedge (e(m * \hat{y}) \neq 0 \wedge em \neq 0 \to R(d^m, c^m, y))\}$$

where

$$d^m = \lambda x.d\ \{m, x\},\ c^m = \lambda x.c\ \{m, x\}.$$

15.3.

A typical application of the full comprehension principle is given by the definition of a component of a point p in a topological space $\langle X, \mathcal{T} \rangle$ as the union of all connected $Y \subseteq X$ such that $p \in Y$. ($Y \subseteq X$ is connected if we cannot find Y_1, Y_2, closed in Y, such that $Y_1 \cup Y_2 = Y$, $Y_1 \cap Y_2 = \emptyset$).

We shall present here a typical (and essential) application of the predicative comprehension principle, which clearly shows the role of the comprehension principle as a creator of new objects.

15.3.1. We write $X \mathcal{R}_1 Y$ if there are (lawlike) mappings ϕ, ϕ', ϕ' bi-unique, such that for some $Z \subseteq \phi X$, $\phi'[\phi[X]] = Y$ ($\phi'[\phi[X]] = \{\phi' y : y \in \phi x\}$).

Let $X \mathcal{R}_2 Y$ mean: there is a (lawlike) mapping ϕ such that $\phi[X] = Y$. (dom ϕ may be properly included in X.)

15.3.2. <u>Theorem</u>. $X \mathcal{R}_2 Y \to X \mathcal{R}_1 Y$ ([T 1967 A], lemma 2.4).

Proof. Let $\phi[X] = Y$, dom $\phi \subseteq X$, where dom $\phi = \{x : \bigvee y(\phi x = y)\}$.
We put for any $x \in X$:

$$\Psi x = \{y : y \in \text{Dom}\ \phi \wedge x \in \text{Dom}\ \phi \wedge \phi x = \phi y\}.$$

If $x \in \text{Dom}\ \phi$, then Ψx is inhabited, and conversely.
We define ϕ' on $\{\Psi y : \bigvee z(z \in \Psi y)\}$ by putting

$$\phi'(\Psi(x)) = \phi x.$$

Clearly

$$\phi'[\{\Psi(y) : \bigvee z(z \in \Psi y)\}] = Y.$$

There remains to be proved that ϕ' is bi-unique.

$$\begin{aligned} \phi'(\Psi(x)) = \phi'(\Psi(y)) &\rightarrow \phi(x) = \phi(y) \\ &\rightarrow \phi^{-1}[\phi(x)] = \phi^{-1}[\phi(y)] \\ &\rightarrow \Psi x = \Psi y. \end{aligned}$$

(We use $\Psi x = \phi^{-1}[\phi(x)] = \{y : y \in \text{Dom}\ \phi \wedge \phi(y) = \phi(x)\}$ for $x \in \text{Dom}\ \phi$).

15.3.3. Remark. The hypothesis $X \mathcal{R}_2 Y$ requires the existence of X, Y, and

$$\{\langle x,y\rangle : x \in X \wedge y \in Y \wedge y = \phi x\}.$$

The proof uses the comprehension principle relative to these species. For another application of weak comprehension, see e.g. [T 1967 A], theorem 4.3.

§ 16. Brouwer's theory of the creative subject

16.1.

In a number of papers published after 1945 (e.g. [Br 1948], p. 1246, [Br 1948 A], [Br 1949], [Br 1949 A]), Brouwer introduced the idea of the creative subject (or the idealized mathematician). This concept gave rise to much discussion and it is likely to do so for some period of time to come. A systematic and coherent theory has not yet been developed, so I am restricted to presenting a few fragments.

The central idea is that of an idealized mathematician (consistent with the subjectivistic viewpoint of intuitionism, we may think of ourselves; or even better, to obtain the required idealization, we may think of ourselves as we should like to be), who performs his mathematical activities in a certain order (you may think of the order given by time). The process of his mathematical activity proceeds in discrete stages. Therefore we introduce a basic notion:

$$\vdash_m A$$

to be read as: "the creative subject has a proof of A at stage m" or better "the creative subject has evidence for A at stage m".

We suppose $\vdash_m A$ to be a decidable relation, i.e.

$$\vdash_m A \vee \neg \vdash_m A.$$

At stage m we know if we have evidence for A or if we do not have evidence for A.

Clearly

$$(\vee m \vdash_m A) \rightarrow A. \qquad (2)$$

"If we have evidence for A at stage m, then we can find a proof of A".

In order to simplify the interpretation we also suppose

$$(\vdash_m A) \wedge (n > m) \rightarrow (\vdash_n A) . \qquad (3)$$

"The evidence at stage m is also contained in all following stages".

If we boldly identify the provable assertions with the assertions for which we can obtain evidence at a certain stage, we also have

$$A \leftrightarrow \vee m(\vdash_m A) \qquad (4)$$

or in combination with (2)

$$\bigvee m(\vdash_m A) \leftrightarrow A \quad . \tag{5}$$

If we want to be more cautious, we may satisfy ourselves with the following assertion instead of (4):

$$A \rightarrow \neg\neg \bigvee m(\vdash_m A) \ . \tag{6}$$

(6) may be read as follows: "I am completely free in making deductions. Hence if there is a proof of A, it is absurd that I would be able to prove that I will never find a proof of A (at no stage will have evidence for A)".

If we want to assert (6), without asserting (4), this means that we do not want to identify all possible constructive proofs with the collection of proofs whose existence becomes evident to me at a certain time (stage).

Let us call in the sequel the theory based on (6) instead of (4) the "weak theory", and the theory based on (4) the "strong theory".

16.2.

In the existing literature, most of the deductions are based on the weak theory. In developing consequences from the weak or the strong theory, I shall try to be cautious and hence proceed more or less axiomatically, in order to show what is actually used in certain deductions.

In the weak system, we can derive the following scheme (called Kripke's scheme in the literature):

$$\bigvee \chi[(\bigwedge x(\chi x = 0) \leftrightarrow \neg A) \wedge (\bigvee x(\chi x \neq 0) \rightarrow A)]. \tag{7}$$

In the strong system, (7) can be strengthened to

$$\bigvee \chi(\bigvee x(\chi x \neq 0) \leftrightarrow A). \tag{8}$$

This is seen by defining χ relative to a given assertion A by:

$$\left.\begin{array}{l}(\vdash_n A) \rightarrow \chi n = 1 \\ (\neg \vdash_n A) \rightarrow \chi n = 0.\end{array}\right\}$$

If we have a definite prescription involving the actions of the creative subject (by means of a relation like $\vdash_n A$) for determining the values of a sequence, we speak of an <u>empirical</u> sequence (as is done e.g. in [M 1968]).

Our idea of lawlike sequence does not exclude empirical sequences, at least not as long as we are willing to consider reference to our own course of activity by means of $\vdash_n$ as "definite".

It is clear however, that e.g. primitive recursive functions are lawlike in a stricter, more objective sense; their values are independent of future decisions about the order in which we want to make deductions.

Let us call a sequence which is given by a complete description not involving $\vdash_n$ a mathematical ([M 1968]) or absolutely lawlike sequence.

If a sequence ξ is defined by a complete description from sequences $x_1, x_2, \ldots,$ without reference to the creative subject, we shall call ξ mathematical or absolutely lawlike in $x_1, x_2, \ldots$.

We shall return to the distinction between empirical and mathematical in 16.8.

It seems to me that the axiom of choice as discussed in section 5 i.e. for an A containing free variables for lawlike objects only)

$$\bigwedge x \bigvee y\, A(x, y) \rightarrow \bigvee a \bigwedge x\, A(x, ax)$$

is really evident only if we do not exclude empirical sequences from the species of lawlike sequences, since it is conceivable that $\bigwedge x \bigvee y\, A(x, y)$ is proved with the use of $\vdash_n$ (even if A itself does not refer to $\vdash_n$).

In order to conclude from a proof of $\bigwedge x \bigvee y\, A(x, y)$ (A not involving $\vdash_n$) to $\bigvee a \bigwedge x\, A(x, ax)$ with a mathematical, we ought to restrict ourselves to arguments not involving $\vdash_n$. (This can be done, I believe, consistently throughout the preceding sections.)

For lawlike sequences, we obtain from (7)

$$\bigwedge a \bigvee b \neg\neg(\bigwedge x(ax = 0) \leftrightarrow \bigvee y(by = 0)) \quad . \qquad (9)$$

If we use (8) instead, we obtain the stronger form:

$$\bigwedge a \bigvee b(\bigwedge x(ax = 0) \leftrightarrow \bigvee y(by = 0)) \quad . \qquad (10)$$

This is seen by applying (7) and (8) respectively to the formula $A \equiv_D \bigwedge x(ax = 0)$.

In virtue of (9), various types of unsolvable problems that are used in intuitionistic "weak" counterexamples collapse, i.e. some of these classes of problems turn out to be essentially equivalent.

A few of these types of problems, some of which we have met before, are represented by the following formulae, which express assertions which we have no hope of proving intuitionistically:

$$\bigwedge b(\bigvee x(bx = 0) \vee \neg\bigvee x(bx = 0)) \tag{11}$$

$$\bigwedge b(\neg\neg\bigvee x(bx = 0) \rightarrow \bigvee x(bx = 0)) \tag{12}$$

$$\bigwedge b(\neg\bigvee x(bx = 0) \vee \neg\neg\bigvee x(bx = 0)) \tag{13}$$

$$\bigwedge a \bigwedge b[\neg(\bigvee x(ax = 0) \wedge \bigvee x(bx = 0)) \rightarrow (\neg\bigvee x(ax = 0) \vee \neg\bigvee x(bx = 0))] . \tag{14}$$

(11) and (13) are restricted forms of the principle of the excluded third and the principle of testability respectively. (12) is an intuitionistic analogue of Markov's principle.

Intuitionistically, (11) is equivalent to

$$\bigwedge b \bigvee c \bigwedge y[(cy = 0 \rightarrow \bigvee x\; b\{x,y\} = 0) \wedge (cy \neq 0 \rightarrow \neg\bigvee x b\{x,y\} = 0)$$

and (14) is equivalent to

$$\bigwedge a \bigwedge b(\bigwedge y \neg\neg(\bigvee x\; a\{x,y\} = 0 \wedge \bigvee x\; b\{x,y\} = 0) \rightarrow$$

$$\bigvee c \bigwedge y[(\bigvee x\; a\{x,y\} = 0 \rightarrow cy = 0) \wedge (\bigvee x\; b\{x,y\} = 0 \rightarrow cy \neq 0)]).$$

The first formula expresses the analgon of the assertion: every recursively enumerable set is recursive, the second one expresses the analogon of: every pair of disjoing r.e. sets can be separated by a recursive set.

For an A not containing non-lawlike variables, (8) simplifies to

$$\bigvee b(A \leftrightarrow \bigvee x(bx = 0)). \tag{15}$$

Since the members of the pairs

$$\neg\neg A \rightarrow A, \neg A \vee A$$

and

$$\neg(A \wedge B) \rightarrow (\neg A \vee \neg B), \neg A \vee \neg\neg A$$

possess equal strength as axiom schemes when added to intuitionistic logic, we see easily that as a consequence of (15), (11) implies $A \vee \neg A$ for all A with lawlike parameters only; likewise (13) implies $\neg A \vee \neg\neg A$ for such A. Therefore (11) $\leftrightarrow$ (12), (13) $\leftrightarrow$ (14).

It is worthwhile knowing however, that (11) $\leftrightarrow$ (12), (13) $\leftrightarrow$ (14) can be proved using (9) only, without further reference to the creative subject.

<u>Proof.</u> (i) (11) → (12) is immediate, even without (9).
Suppose (12), and take any b. According to (9) we can find a b such that

$$\neg\neg(\neg Vy(by = 0) \leftrightarrow Vy(cy = 0)) .$$

So

$$\neg Vy(by = 0) \leftrightarrow \neg\neg Vy(cy = 0).$$

Since $\neg\neg(Vy(by = 0) \vee \neg Vy(by = 0))$, it follows that

$$\neg\neg(Vy(by = 0) \vee \neg\neg Vz(cz = 0)) .$$

$P \vee \neg\neg Q \to \neg\neg(P \vee Q)$ is a theorem of intuitionistic propositional logic, hence

$$\neg\neg(Vy(by = 0)) \vee Vz((cz = 0)),$$

so

$$\neg\neg Vx(bx.cx = 0).$$

Applying (12), we conclude to $Vx(bx.cx = 0)$, hence $Vx(bx = 0) \vee Vx(cx = 0)$, which is equivalent to

$$Vx(bx = 0) \vee \neg Vx(bx = 0).$$

(ii). Suppose (13). For a, b **such** that $\neg(Vx(ax = 0) \wedge Vx(bx = 0))$ we conclude to

$$\neg\neg Vx(ax = 0) \leftrightarrow Vx(bx = 0)$$
$$\neg Vx(ax = 0) \leftrightarrow \neg\neg Vx(bx = 0).$$

Then by (13) $\neg Vx(ax = 0) \vee \neg\neg Vx(ax = 0)$.

Conversely, suppose (14).
Take any b, and let a be such that

$$\neg\neg(\Lambda x(bx \neq 0) \leftrightarrow Vy(ay = 0)).$$

Then $\neg\neg \Lambda x(bx \neq 0) \leftrightarrow \neg\neg Vy(ay = 0))$
and $\neg \Lambda x(bx \neq 0) \leftrightarrow \neg Vy(ay = 0)$.

From (14) and the fact that $\neg(Vx(bx = 0) \wedge Vy(ay = 0))$ it follows that $\neg Vx(ax = 0) \vee \neg Vx(bx = 0)$ therefore we may conclude to $\neg Vx(bx = 0) \vee \neg\neg Vx(bx = 0)$.

Now we shall proceed with somewhat more interesting theorems.

16.3.

Theorem. We can prove in the weak system (in fact, using (9) only):

$$\neg\bigvee b\bigwedge a\bigvee x\bigwedge z[\bigvee y\ a\{z,y\} = 0 \leftrightarrow \bigvee u\ b\{x,\{z,u\}\} = 0]\ . \qquad (16)$$

This result has been called a refutation of "Church's thesis", but since empirical sequences are rather far removed from the idea of "mechanically computable" functions, it is perhaps better to describe it as a non-enumerability result, as Myhill proposed ([M 1967]).

Proof. Suppose

$$\bigwedge a\bigvee x\bigwedge z\{\bigvee y\ a\{z,y\} = 0 \leftrightarrow \bigvee u\ b\{x,\ \{z,u\}\} = 0\}\ . \qquad (17)$$

We remark that

$$\neg\bigvee u\ b\{x,\{x,u\}\} = 0 \leftrightarrow \bigwedge u(1 \dot{-} b\{x,\{x,u\}\} = 0) \leftrightarrow \bigwedge u(b\{x,\{x,u\}\} \neq 0)\ .$$

Now we can find (by (9)) a c such that

$$\neg\neg(\bigwedge u\ b\{x,\{x,u\}\} \neq 0 \leftrightarrow \bigvee v\ c\{x,v\} = 0)\ . \qquad (18)$$

In virtue of our hypothesis (17), there exists an x_0 such that

$$\bigwedge x(\bigvee v\ c\{x,v\} = 0 \leftrightarrow \bigvee w\ b\{x_0,\{x,w\}\} = 0\ . \qquad (19)$$

So we obtain

$$\begin{aligned}\neg\neg\bigvee w\ b\{x_0,\{x_0,w\}\} = 0 &\leftrightarrow \neg\neg\bigvee v\ c\{x_0,v\} = 0 \quad \text{(from (19))}\\ &\leftrightarrow \neg\neg\bigwedge u\ b\{x_0,\{x_0,v\}\} \neq 0 \quad \text{(from (18))}\\ &\leftrightarrow \bigwedge u\ b\{x_0,\{x_0,v\}\} \neq 0\\ &\leftrightarrow \neg\bigvee w\ b\{x_0,\{x_0,w\}\} = 0\ :\end{aligned}$$

contradiction. This disproves (17).

16.4.

Theorem ([M 1967]). There exists an extensional predicate B such that $\bigwedge\chi\bigvee\chi'\ B(\chi,\chi')$, but for no continuous functional Γ of type $((N)N)(N)N$ $\bigwedge\chi B(\chi,\Gamma\chi)$.

Proof. Apply Kripke's scheme (7) to

$$A\chi \equiv_D \bigvee x\bigwedge y > x(\chi y = 0).$$

Then we obtain

$$\left.\begin{array}{l}\bigwedge\chi\bigvee\chi'[(\neg\bigvee x\bigwedge y > x(\chi y = 0) \leftrightarrow \bigwedge x(\chi' x = 0))\wedge \\ \quad \wedge(\bigvee x\chi' x \neq 0 \to \bigvee x\bigwedge y > 0(\chi y = 0))] .\end{array}\right\} \qquad (20)$$

Let us denote the part of (20) within the square brackets by $B(\chi,\chi')$, and suppose that for some continuous Γ

$$\bigwedge\chi B(\chi,\Gamma\chi).$$

In case $(\Gamma\chi)x \neq 0$, it follows that $\bigvee x\bigwedge y > x(\chi y = 0)$.

Since Γ is a continuous functional, there exists an initial segment $\bar{\chi}y$ of χ such that

$$\bigwedge\chi''(\bar{\chi}y = \bar{\chi}''y \to (\Gamma\chi)x = (\Gamma\chi'')x)$$

and therefore

$$\bigwedge\chi''(\bar{\chi}y = \bar{\chi}''y \to \bigvee x\bigwedge z > x(\chi''z = 0)).$$

This is obviously false; we only have to take for χ'' the function such that

$$\bar{\chi}''y = \bar{\chi}y,\ \chi''(y+z) = 1 \quad \text{for all } z.$$

Hence $(\Gamma\chi)x = 0$ leads to a contradiction; therefore $\bigwedge\chi\bigwedge x(\Gamma\chi)x = 0$, i.e. Γ is the zero functional. As a consequence $\bigwedge\chi\neg\bigvee x\bigwedge y > x(\chi y = 0)$, which is a plain contradiction, since the zero function provides a counterexample.

16.5.

16.5.1. <u>Theorem</u>. $\neg\bigwedge\chi(\neg\neg\bigvee x(\chi x = 0) \to \bigvee x(\chi x = 0))$ ([Kr 1967], p. 160).

<u>Proof</u>. We apply (7) to $A\chi' \equiv_D \bigvee x(\chi' x = 0) \vee \neg\bigvee x(\chi' x = 0)$:

$$\bigvee\chi[(\neg\neg\bigvee x(\chi x = 0) \leftrightarrow \neg\neg A\chi')\wedge\bigvee x(\chi x = 0) \to A\chi')].$$

Take for χ' any choice sequence β.

Since $\neg\neg A\beta$ holds ($A\beta$ is an application of the principle of the excluded third) we find a χ such that $\neg\neg\bigvee x(\chi x = 0)$, hence if $\bigwedge\chi(\neg\neg\bigvee x(\chi x = 0) \to \bigvee x(\chi x = 0))$ would hold, then $\bigvee x(\chi x = 0)$, hence $A\beta$. $\bigwedge\beta A\beta$ is contradictory (compare 9.10 (I)); so the assertion of the theorem follows.

16.5.2. <u>Remark</u>. At first sight, 16.5.1 is just weaker than the result 9.10 (V) in section 9; but actually, we have proved more, we have shown:

$$\neg\bigwedge\chi\in\mathfrak{A}\,(\neg\neg\bigvee x(\chi x = 0) \to \bigvee x(\chi x = 0))$$

where $\mathfrak{A}$ is the class of sequences obtained from choice sequences by a lawlike (but not necessarily extensional) operation.

16.6.

Brouwer gave ([Br 1948 A]) a simple (weak) counterexample of a lawlike real number x such that $x \neq 0$, while we have no proof of $x \# 0$. To do this, he takes an assertion A which has not been tested up till now (that is, we do not know whether $\neg A$ or $\neg\neg A$), and defines x by a real number generator $\langle r_n \rangle_n$:

$$\neg \vdash_n (\neg A \vee \neg\neg A) \rightarrow r_n = 0$$
$$\vdash_m (\neg A \vee \neg\neg A) \wedge \neg \vdash_{m-1} (\neg A \vee \neg\neg A) \wedge n > m \rightarrow r_n = 2^{-m}.$$

Then one verifies easily that in the weak or strong system $x \neq 0$; but $x \# 0$ would imply $\neg A \vee \neg\neg A$; so we do not have a proof of $x \# 0$.

We can make use of the equivalence of (11) and (12) relative to the theory of the creative subject, in order to obtain another formulation of this result without further reference to the creative subject.

Let x_b denote a real number with a real number generator $\langle r_n \rangle_n$, defined w.r.t. a lawlike function b by

$$\left.\begin{array}{l} \neg \bigvee m \leq n(bm \neq 0) \rightarrow r_n = 0 \\ (m \leq n \wedge bm \neq 0 \wedge \bigwedge n' < m(bn' = 0)) \rightarrow r_n = 2^{-m} \end{array}\right\}$$

then (12) is equivalent with $\bigwedge b(x_b \neq 0 \rightarrow x_b \# 0)$; but on the other hand (12) implies (11), therefore

$$\bigwedge x(x \neq 0 \rightarrow x \# 0) \rightarrow (11) .$$

(11) implies in turn $\bigwedge x(x = 0 \vee x \# 0)$. For let for any x $\langle r_n \rangle_n$ be such that $\bigwedge n |x - r_n| < 2^{-n-1}$. Then $\bigwedge n(r_n - 2^{-n} < x < r_n + 2^{-n})$, $\langle r_n - 2^{-n} \rangle_n \in x$, $\langle r_n + 2^{-n} \rangle_n \in x$. Hence $x \# 0 \leftrightarrow \bigvee n(r_n - 2^{-n} > 0 \vee r_n + 2^{-n} < 0)$, and if we put $bn = 0 \leftrightarrow r_n - 2^{-n} > 0 \vee r_n + 2^{-n} < 0$, we see that (11) implies $\bigwedge x(x \# 0 \vee x = 0)$. Therefore for lawlike x

$$\bigwedge x(x \neq 0 \rightarrow x \# 0) \rightarrow \bigwedge x(x = 0 \vee x \# 0) .$$

Some other examples of theorems to which we can find weak counterexamples of the same kind:

$$\bigwedge x(x \neq 0 \rightarrow x > 0 \vee x < 0),$$
$$\bigwedge y \bigwedge z(y \neq 0 \wedge z \neq 0 \rightarrow \bigvee x(yx + z = 0)) .$$

If we sufficiently enlarge our notion of real number, we can give mathematical disproofs by means of theorems like 16.5.1 , e.g.

$$\neg\bigwedge x(x \neq 0 \rightarrow x \# 0)$$

is provable if the class of reals is sufficiently large (roughly speaking: take all reals obtained from real number generators $\langle r_{\chi n}\rangle_n$, $\langle r_n\rangle_n$ being a standard enumeration of the rationals, $\chi \in \mathfrak{A}$ as indicated in remark 16.5.2).

16.7.

16.7.1. Lemma. Let a, b be two lawlike sequences such that there exist bi-unique mappings ψ, ϕ satisfying

$$\phi[\text{Range }(a)] \subseteq \text{Range }(b),\ \psi[\text{Range }(b)] \subseteq \text{Range }(a).$$

Then there exists a bi-unique ξ defined on Range (a) such that

$$\xi[\text{Range }(a)] = \text{Range }(b)\ .$$

(Or if ξ is taken to be defined on Range (a) only, Range (ξ) = Range (b).)

Proof. We define a' such that $a'x = y \leftrightarrow ax$ is the $(y+1)^{\text{th}}$ "new" number generated by a. Or explicitly:

$$a'x = y \leftrightarrow \bigvee z < x(a(z+1) = ax \wedge ax \notin \{a0,\dots, az\} \wedge$$
$$\{a0,\dots,az\} \text{ contains } y \text{ different elements}) \vee (ax = a0 \wedge y = 0)$$

b' is defined likewise from b.

Now we define ξ as follows. Consider ax, and let $a'x = y$. Then we can find a bz such that $b'z = y$, because $\{\phi(a\ 0),\dots, \phi(ax)\}$ must contain at least y different elements. Now we put $\xi ax = bz$.

We then use the existence of ψ to demonstrate $\bigwedge z \bigvee x(\xi ax = bz)$.

16.7.2. Lemma. Let X be an inhabited species of natural numbers. Then there is a lawlike sequence a such that

$$\text{Range }(a) = X.$$

Proof. For sake of simplicity, let us suppose $0 \in X$, $\vdash_0 0 \in X$.
Then we put $a0 = 0$. a is further defined by:

$$\neg \vdash_m z \in X \rightarrow a\{m,z\} = a0$$
$$\vdash_m z \in X \rightarrow a\{m,z\} = z.$$

Now one verifies easily that on account of

$$x \in X \leftrightarrow \bigvee m(\vdash_m x \in X)$$

(an application of (6)) we have Range (a) = X.

Remark. The proof as described above proceeds directly from the axioms for the creative subject, but we could have obtained the result also from the specialization of (8) to species:

$$\bigwedge x \bigvee a [\bigvee y(ay = 0) \leftrightarrow x \in X];$$

hence $\bigwedge x[\bigvee y(b\{x,y\} = 0 \leftrightarrow x \in X]$ for a certain b, and since X is inhabited, we may construct in the usual way a constructive function which enumerates $\{x : \bigvee y\ b\{x,y\} = 0\}$.

16.7.3. Theorem. Let X, Y be inhabited species of natural numbers, and let ϕ, ψ be bi-unique mappings, $\phi \in (X)Y$, $\psi \in (Y)X$. Then there exists a bi-unique mapping $\xi \in (X)Y$ such that $\xi[X] = Y$ (Intuitionistic analogue of Cantor-Bernstein).

Proof. Immediate by 16.7.1, 16.7.2.

16.7.4. Remark. The requirement that X, Y are inhabited can be omitted; the proof is slightly more involved in that case.

We would have had no hope of obtaining this result if we had required ξ to be mathematical in ϕ, ψ. This is demonstrated by the following counterexample: take $X = N$, $Y = \{2n+1 : n \in N\} \cup \{m : \neg \bigvee x \Pi_m(x)\}$, $\phi n = 2n + 1$, $\psi n = n$.

It is instructive to compare 16.7.3 with the following theorem in the literature, which can be proved without reference to the creative subject:

16.7.5. Theorem. Let X, Y be inhabited subspecies of the natural numbers, and let ϕ, ψ be bi-unique mappings of X into Y, Y into X respectively, such that $\phi[X]$ is detachable in Y, $\psi[Y]$ detachable in X. Then there exists a bi-unique mapping ξ from X onto Y.

Proof. (formulated with "recursive" instead of "lawlike")
[D, M 1960] theorem 13(b) or [Rog 1967] § 7.4, theorem VI.

As it stands, 16.7.5 is just weaker than 16.7.4. However, the existing proofs of 16.7.5 contain more information than is contained in 16.7.5, since they do not refer to the creative subject. Hence we may strengthen 16.7.5 to:

16.7.6. Theorem. Under the conditions of 16.7.5, there exists a bi-unique mapping ξ from X onto Y, ξ mathematical in ϕ, ψ.

The formulation of many results in the preceding sections may be strengthened in the same manner.

Our intuitive insight that there is an essential difference between empirical and mathematical sequences, can be expressed to a certain extent in the theory of the creative subject.

Take the following property as an example. Let a_A be an empirical sequence associated with a definite assertion A (i.e. A does not contain non-lawlike variables), a_A defined by

$$\vdash_x \neg A \to a_A x = 1, \; \vdash_x A \to a_A x = 2, \; \neg \vdash_x (A \vee \neg A) \to a_A x = 0.$$

Let a_{math} be a variable for mathematical sequences.
Then

$$(\neg \vdash_x (A \vee \neg A)) \to \neg \vdash_x (a_{math} = a_A)$$

expresses a difference between mathematical and empirical sequences.

16.8.

Let me finish with the indication of some points in the theory of the creative subject in need of further clarification. Intentionally, I used the phrase "the creative subject has evidence to assert A at stage m" in preference to saying downrightly "the creative subject proves A at stage m".

The second formulation suggests the creative subject as making his conclusions one by one, so we might ask ourselves if we could not assert that with a sufficient "refinement" in the distinction of the stages, the creative subject draws one conclusion at a time. This supposition can be expressed axiomatically by introducing the constant $A^{(m)}$ (denoting the new conclusion of stage m) by the statement:

$$\vdash_m A^{(m)} \wedge \bigwedge n(n < m \to \neg \vdash_n A^{(m)}) \wedge$$
$$\{(\vdash_m B \wedge \bigwedge n(n < m \to \neg \vdash_n B)) \to (B \leftrightarrow A^{(m)})\}.$$

In some respects this looks plausible; however, in case we accept this axiom, $\vdash_m \bigwedge x\, Ax$ would oblige us beforehand to make in the future all deductions $A0, A1, A2, A3, \ldots$ explicitly, since

$$\bigwedge x\, Ax \leftrightarrow \bigwedge x \bigvee m \vdash_m Ax,$$

and this seems to be a rather unnatural restraint on the future activity of the creative subject.
But even worse: our new axiom leads to a downright contradiction.

For since it is natural to assume that we know when a conclusion has the form "a is a lawlike sequence" we have:

$A^{(m)}$ is a conclusion of the form "a is a lawlike sequence" or

$A^{(m)}$ is a conclusion of another kind.

Then it is possible for us to enumerate the $A^{(m)}$ of the form "a is lawlike"; let $A^{(bx)}$ be the xth conclusion of this form, stating "a_x is a lawlike sequence". Then

$$\Lambda x V a (A^{(bx)} \equiv \text{"}a \text{ is a lawlike sequence"})$$

and so we conclude to the existence of a b' such that

$$b'\{x, y\} = a_x(y) .$$

Intuitively $c = \lambda x . b'\{x,x\} + 1$ is a lawlike sequence, but then we ought to be able to indicate a $z \in N$ such that

$$A^{(bz)} \equiv \text{"}c \text{ is a lawlike sequence"}$$

which implies:

$$\Lambda x (b'\{x,x\} + 1 = b'\{z,x\})$$

which is contradictory. Clearly, the axiom of "one-conclusion at a time" makes the highly impredicative character of the theory of the creative subject acutely paradoxical.

Therefore we must leave open the possibility of infinitely many assertions A for which $\vdash_m A$, and hence the cautious formulation "having evidence for A at stage m". Example: if we have proved $\Lambda x A x$ at stage m, we also have $\vdash_m Ax$ for any special number x.

But since we also required $\vdash_m A \vee \neg \vdash_m A$, we are left with the quite genuine problem of finding a satisfactory interpretation of "evidence for A at stage m". We might try something like "trivially deducible from the available data at stage m", in the same manner as An is trivially deducible from $\Lambda x A x$, but once again we are left with the problem of a general interpretation of "trivially deducible" which ensures decidability at any stage.

Another possibility which is suggested by the "paradox" is, that the primary cause of our trouble is not so much in the assumption of the "one-conclusion-at-a-time" axiom, as well as in the unrestricted possibilities for self-reference implicit in our axioms for the creative subject. Specifically: although assertions like $\vdash_n A$ and $\vdash_m(\vdash_n A)$ do belong to different "levels of self-reflection" (self-reflection = mathematical consideration of the course of our own mathematical activities) we did not distinguish between them in this respect.

So the following possible alternative for a theory of the creative subject suggests itself. To each mathematical assertion and construction we suppose a level (of self-reflection) to be assigned. We restrict ourselves to assertions and constructions of finite level; the unions of all assertions and constructions of finite levels form the assertions and constructions of level ω. Assertions which may be understood or constructions which can be carried out without reference to $\vdash_n$ are said to belong to level zero.

Assertions which are described using $\vdash_n A$ for A of level p and constructions of level p, are said to belong to level $p+1$. Likewise, constructions defined relative to $\vdash_n A$ for A of level p are said to be of level $p+1$.
We make further the following basic <u>general assumption</u>:
"When an assertion of level p has a proof, it has a proof of level p". Now, if we use $a^{(p)}$ to denote a constructive function of level p, we can assert (instead of (10))

$$\bigwedge a^{(p)} \bigvee b^{(p+1)} (\bigwedge x(ax = 0) \leftrightarrow \bigvee y(by = 0)).$$

So if we use a, b as variables for constructive functions of level ω, we still get at level ω

$$\bigwedge a \bigvee b (\bigwedge x(ax = 0) \leftrightarrow \bigvee y(by = 0)).$$

On the other hand, our paradox cannot be derived anymore.

I feel this approach deserves further investigation. Another problem we did not touch upon, is the combination of the idea of incomplete (choice-) objects in connection with the theory of the creative subject. For example, how are the "stages" of the theory of lawless and choice sequences to be related with the "stages" in the theory of the creative subject?

In conclusion we might say that the theory of the creative subject is provocative, attractive, and dangerous; it represents the extreme consequences of intuitionistic subjectivism; undoubtedly it deserves further study, precisely for this reason.

16.9.

<u>References</u>. An easily accessible exposition of Brouwer's counterexamples involving the creative subject is to be found in [H 1966], 8.8.1 (pp. 115-117). Kreisel ([Kr 1967]p. 160) introduced the relation $\vdash_m$ (there appearing as $\Sigma\vdash_m$). A further discussion in connection with analysis is to be found in [M 1967] and [M 1968].

§ 17. Bibliography

This list only contains items referred to in the paper. For a more complete bibliography concerning intuitionism, see [H 1955], [H 1966].

Br 1920 L.E.J. BROUWER, Besitzt jede reelle Zahl eine Dezimalbruchentwickelung? Proc. Akad. Amsterdam 23, pp. 955 - 965.

Br 1924 --, Zur Begründung der intuitionistischen Mathematik I. Mathematische Annalen 93, pp. 244 - 258.

Br 1925 --, Zur Begründung der intuitionistischen Mathematik II. Mathematische Annalen 95, pp. 453 - 473.

Br 1926 --, Zur Begründung der intuitionistischen Mathematik III. Mathematische Annalen 96, pp. 451 - 489.

Br 1926A --, Über Defintionsbereiche von Funktionen. Mathematische Annalen 97, pp. 60 - 76.

Br 1927 --, Virtuelle Ordnung und unerweiterbare Ordnung. Journal für reine und angewandte Mathematik 157, pp. 255 - 258.

Br 1942 --, Zum freien Werden von Mengen und Funktionen. Proc. Akad. Amsterdam 45, pp. 322 - 323 = Indagationes Math. 4, pp. 107-10

Br 1942A --, Beweis dass der Begriff der Menge höherer Ordnung nicht als Grundbegriff der intuitionistischen Mathematik in Betracht kommt. Proc. Akad. Amsterdam 45, pp. 791 - 793 = Indagationes Math. 4, pp. 274-27

Br 1948 --, Consciousness, philosophy and mathematics. Proc. Xth international Congress Philosophy (Amsterdam 1948), pp. 1235 - 1249.

Br 1948A --, Essentieel negatieve eigenschappen. Proc. Akad. Amsterdam 51, pp. 963 - 965 = Indagationes Math. 10, pp. 322-3

Br 1949 --, De non-equivalentie van de constructieve en de negatieve orderelatie in het continuum. Proc. Akad. Amsterdam 52, pp. 122 - 125 = Indagationes Math. 11, pp. 37-40

Br 1949A --, Contradictoriteit der elementaire meetkunde.
Proc. Akad. Amsterdam 52, pp. 315 - 316 = Indagationes Math. 11, pp. 89-90.

Br 1952 --, Historical background, principles and methods of intuitionism.
South African Journal of Science 49, pp. 139 - 146.

D, M 1960 J.C.E. DEKKER and J. MYHILL, Recursive equivalence types.
Berkeley and Los Angeles 1960.

F 1936 H. FREUDENTHAL, Zur intuitionistischen Deutung logischer Formeln.
Compositio Math. 4, pp. 112 - 116.

G 1958 K. GÖDEL, Über eine bisher noch nicht benutzte Erweiterung des finiten Standpunktes.
Dialectica 12, pp. 280 - 287.

Gn 1968 N. GOODMAN, Intuitionistic arithmetic as a theory of constructions.
Ph.D. thesis, Stanford University.

Gn --, A theory of constructions equivalent to arithmetic. To appear.

H 1955 A. HEYTING, Les fondements des mathématiques. Intuitionnisme. Théorie de la démonstration. Paris-Louvain 1955.

H 1959 --, Infinitistic methods from a finitist point of view.
Proc. Symp. Warszawa 1959, Warszawa 1959, pp. 185 - 192.

H 1966 --, Intuitionism, an introduction. Second, revised edition.
Amsterdam 1966.

Ho 1968 W.A. HOWARD, Functional interpretation of bar induction by bar recursion.
Compositio Math. 20, pp. 107 - 124.

Ho,K 1966 W.A. HOWARD and G. KREISEL, Transfinite induction and bar induction of types zero and one, and the role of continuity in intuitionistic analysis.
J. of Symbolic Logic 31, pp. 325 - 358.

K 1952 S.C. KLEENE, Introduction to metamathematics.
Amsterdam, Groningen, New York, Toronto 1952.

K,V 1965 S.C. KLEENE and R.E. VESLEY, Foundations of intuitionistic mathematics.
Amsterdam 1965.

Kr 1958 G. KREISEL, A remark on free choice sequences and the topological completeness proofs.
J. of Symbolic Logic 23, pp. 369 - 388.

Kr 1958A --, Elementary completeness properties of intuitionistic logic with a note on negations of prenex formulae.
J. of Symbolic Logic 23, pp. 317 - 330.

Kr 1962 --, On weak completeness of intuitionistic predicate logic.
J. of Symbolic Logic 27, pp. 139 - 158.

Kr 1965 --, Mathematical logic. pp. 95 - 195 in:
Lectures on modern mathematics, vol. 3, ed. T.L. Saaty, New York 1965.

Kr 1967 --, Informal rigour and completeness proofs. pp. 138 - 186 in:
Problems in the philosophy of mathematics, ed. I. Lakatos, Amsterdam 1967.

Kr 1968 --, Lawless sequences of natural numbers.
Compositio Math. 20, pp. 222 - 248.

Kr 1968A --, Functions, ordinals, species, pp. 145 - 159 in:
Proc. 3rd intern. congress for logic, methodology and philosophy of science, ed. B. van Rootselaar and J.F. Staal, Amsterdam 1968.

Kr,T G. KREISEL and A.S. TROELSTRA, A formal system for intuitionistic analysis (in preparation).

M 1967 J. MYHILL, Notes towards a formalization of intuitionistic analysis.
Logique et Analyse 35, pp. 280 - 297.

M 1968 --, Formal systems of intuitionistic analysis I, pp. 161 - 178 in:
Proc. 3rd intern. congress for logic, methodology and philosophy of science, ed. B. van Rootselaar and J.F. Staal, Amsterdam 1968.

N 1966 T. NAGASHIMA, An extension of the Craig-Schütte interpolation theorem.
Annals of the Japan Association for Philosophy of Science 3, pp. 12 - 18.

S 1962 K. SCHÜTTE, Der Interpolationssatz der intuitionistischen Prädikatenlogik.
Mathematische Annalen 148, pp. 192 - 200.

Sp 1962 C. SPECTOR, Provably recursive functionals of analysis; a consistency proof of analysis by an extension of principles formulated in current intuitionistic mathematics. pp. 1 - 27 in:
Recursive function theory, Proc. Symp. Pure Mathematics (1962).

T 1966 A.S. TROELSTRA, Intuitionistic general topology. Thesis. Amsterdam 1966.

T 1967 --, Intuitionistic continuity.
Nieuw Archief voor Wiskunde (3) 15, pp. 2 - 6.

T 1967A --, Finite and infinite in intuitionistic mathematics.
Compositio Math. 18, pp. 94 - 116.

T 1968 --, The theory of choice sequences, pp. 201 - 223 in:
Proc. 3rd intern. congress for logic, methodology and philosophy of science, ed. B. van Rootselaar and J.F. Staal, Amsterdam 1968.

T 1968A --, One-point compactifications of intuitionistic locally compact spaces.
Fundamenta Mathematicae 57, pp. 75 - 93.

T 1968B --, The use of "Brouwer's principle" in intuitionistic topology, pp. 289 - 298 in:
Contributions to Mathematical Logic. ed. by K. Schütte, Amsterdam 1968.

Rog 1967 Hartley ROGERS Jr., Theory of recursive functions and effective computability, New York etc. 1967.

R 1960 B. van ROOTSELAAR, On intuitionistic difference relations.
Proc. Akad. Amsterdam A 63 = Indagationes Math. 22, pp. 316 - 322.

R 1963 --, Corrections to the paper "On intuitionistic difference relations".
Proc. Akad. Amsterdam A 66 = Indagationes Math. 25, pp. 132 - 133.

Offsetdruck: Julius Beltz, Weinheim/Bergstr.

Lecture Notes in Mathematics

Bitte wenden / Continued